Milton Harris: Chemist, Innovator, and Entrepreneur

Milton Harris: Chemist, Innovator, and Entrepreneur

Miklos M. Breuer, EDITOR

Published from papers presented
at the Milton Harris 75th Birthday Symposium
held April 27–28, 1981 in Washington, DC

Sponsored by:

The American Chemical Society
The American Institute of Chemists
The Fiber Society
Gillette Research Institute
Textile Research Institute

AMERICAN CHEMICAL SOCIETY
WASHINGTON, D. C. 1982

Library of Congress Cataloging in Publication Data

Milton Harris, chemist, innovator and entrepreneur.

"Papers presented at the Milton Harris 75th Birthday Symposium held April 27–28, 1981 in Washington, DC, sponsored by The American Chemical Society . . . [et al.]."

Bibliography: p.
Includes index.

1. Polymers and polymerization—Congresses. 2. Textile fibers—Congresses. 3. Science and state—United States—Congresses. 4. Harris, Milton, 1906– . —Congresses.
I. Breuer, Miklos M., 1927– . II. American Chemical Society.

QD380.M55 1982 677'.02832 82-13926
ISBN 0-8412-0740-2 AACR2 1-169
1982

PRINTED IN THE UNITED STATES OF AMERICA

CONTENTS

EDITOR'S PREFACE

Celebrating a 75th birthday is always a joyous event. It is a particularly pleasant occassion when "the Birthday-Child" is an eminent scientist, a mentor, and an old friend.

When the idea of organizing a symposium honoring Dr. Milton Harris' 75th birthday first emerged during an after dinner conversation with William Bailey, I did not fully realize the breadth and ramifications of Milton Harris' influence on the scientific and technical life of our generation of scientists and engineers. Therefore, the task of covering the areas to which Dr. Harris has made important contributions, appeared at first quite formidable. However, we were relieved when we learned that practically all the speakers we approached were more than willing to give a lecture in honor of Milton Harris.

To complete our celebration, our committee felt that a lasting record of the Symposium in a form of a "Festschrift" should also be published. The present volume represents, then, our tribute to Dr. Milton Harris' scientific and technical achievements.

I should like to take this opportunity to thank all my colleagues on the Symposium Committee and, in particular, William Bailey, our program chairman, for the work they did to assemble this splendid lecture program. Our thanks also go to the speakers who, despite many pressing commitments, took the time and prepared their lectures for print.

Personally, I also should like to thank members of the ACS Meeting Organization, ACS Books Department Staff, and especially Rod Hader, the secretary of the Society, for all the labor they invested in the success of the Symposium and this book.

Finally, thank-you Milton for giving us the occasion to be merry! Many happy returns!

Miklos M. Breuer

The Gillette Company
Boston, Ma.
July 8, 1982

The friends and colleagues of

Milton Harris

cognizant of his many contributions
to the advancement of chemistry
in all its branches,
noting especially his work in
natural and synthetic polymers,
fiber and textile science,
and chemical information systems,
and grateful for his leadership in
strengthening scientific and technical societies
while encouraging more productive interactions
between the scientific community and government,
now join in expressing their esteem
through the

Milton Harris 75th Birthday Symposium

held in his honor on
April 27 and 28, 1981
in Washington, D.C.

Presented on April 27, 1981

PREFACE

The idea of a 75th birthday symposium of scientific contributions by outstanding scientists to honor the long, distinguished career of Milton Harris originated with Miklos Breuer of the Gillette Research Institute. A committee was formed to organize the festivities and comprised Miklos Breuer; Carolyn Damon, American Institute of Chemists; Norman Hollies, The Fiber Society; Ludwig Rebenfeld, Textile Research Institute; Rodney Hader and Raymond Mariella, American Chemical Society; and me, William Bailey, University of Maryland and then Chairman of the Board of Directors of the American Chemical Society. The committee appointed me as program chairman. The concept evolved of a two-day celebration including a symposium of scientific presentations, a birthday banquet, and a panel discussion with Milton. The diversity in the financial sponsors attest to Milton's ubiquity in the sciences: American Chemical Society and Gillette Research Institute (the primary contributors); Celanese Corporation; Chesebrough-Pond's, Inc.; Colgate-Palmolive Company; Clairol, Incorporated; The Clorox Company; The Dow Chemical Company; E.I. duPont de Nemours & Company, Inc.; International Wool Secretariat; Johnson & Johnson; Monsanto Company; J. P. Stevens & Company, Inc.; and Union Carbide Corporation.

Milton Harris' spheres of influence extend through many areas of science as well as into scientific organizations, government agencies, and public activities. Therefore, the symposium honoring Milton Harris could only be representative of his achievements; to present a complete recital of his contributions would take much longer than two days. Thus, the committee chose polymer and cellulose chemistry, keratin and wool science, chemical information science, and science policy as fields for review in this symposium, as it was felt that Milton's contributions to these areas were especially significant.

On Monday evening Milton was the guest of honor at a birthday banquet. A scroll was presented to him. (A copy of the scroll is on page viii.) Representatives of organizations with which Milton was associated in his career conveyed birthday greetings and shared their recollections: Ernest Ambler, National Bureau of Standards; Alfred Brown, Harris Research Laboratories; John Bush, the Gillette Company; Ludwig Rebenfeld; Janet Berry, American Institute of Chemists; Bradford Stanerson, American Chemical Society; John McShefferty, Gillette Research Institute; Darrah Thomas, Oregon State University; and Bingham Humphrey,

Yale University. I was honored to be the master of ceremonies for the evening. Special guests were Milton's wife, Carolyn; his son and daughter-in-law, John and Nancy, and their son, Barney; his daughter and son-in-law, Barbara and Donald Rothenberg.

The panel discussion on Tuesday morning was planned to induce Milton to reminisce about his career. The panel, consisting of Alfred Brown (moderator), Bradford Stanerson, Arnold Sookne of Burlington Industries, Ludwig Rebenfeld, and me, encouraged Milton by asking questions and making comments. A transcription of the panel interview is included in this book.

Special thanks should go to the eminent contributors of the symposium. All of the speakers are very busy scientists and some of them came from great distances to honor Milton. Their presentations, printed in this monograph, make a fitting tribute to Milton's scientific achievements.

On behalf of the committee, I want to thank all those who helped make the Milton Harris 75th birthday symposium so memorable, and to warn them that we will need their help again—when it comes time for the Milton Harris 100th birthday symposium!

William J. Bailey

Chairman, Board of Directors,
American Chemical Society, 1981
May 1982

1

Milton Harris Meets the Panel

A Panel Discussion of Reminiscences

Alfred E. Brown (Moderator), William J. Bailey, Dale Baker, C. S. Marvel, Arnold M. Sookne, Bradford R. Stanerson, and Edward A. Calnan

A.E. Brown: Last night at the banquet many people said all of those wonderful things about Milton and he had only a few minutes to respond. Today, the five of us on this panel will say very little. We will try with our questions to get Milton to do most of the talking.

Milt, I'm going to ask you a question that I always ask famous chemists. How did you ever become a chemist?

M. Harris: I was raised in Portland, Oregon. The whole state of Oregon had less than 500,000 people, so Portland was the biggest city; I don't know how large it was. But as a youngster, 12 or 13 years of age, my parents had a mom-pop grocery store, opened from 7 in the morning to 10 at night, 7 days a week. My principal occupation, I delivered groceries. Along the way there was a city library, a 3-story building a square block in dimension. And whenever I wasn't delivering groceries, I was in the library. I was browsing around, I don't understand how or why, looking at technical books. I remember very vividly a book on how they built the Panama Canal and the first New York skyscraper. I don't remember looking at chemical books at this time. But, someplace along the line I remember Slaussen's "Creative Chemistry" but it must have been later. But, at this age of 12 to 13 I built crystal sets. This became my first business enterprise. The milkman, who delivered milk to the store, had me make one for him. He gave me two dollars for it. But it didn't make any difference. Now, in high school we had a science course and it turned out to be chemistry. The teacher was a little short fellow who stood on the little box behind the lectern. He got me excited about chemistry. At that point in my history, I was open to anything technical. If he had been a physicist or an engineer I'd have been a physicist or an engineer. There's a moral here though. Today I'm involved with a number of universities where we talk about not getting enough students into chemistry. I feel very strongly that to deal with this problem we'd better look at our

8412-0740-2/82-0001$06.00/0

The panel. Dr. Harris (center) with moderator A. E. Brown (left) and William J. Bailey (right).

early educational programs. I know for one thing as I look back that chemists have become a little elitist; they like to express theories. As we learned chemistry in the 1920's and I think Butch Hanford would agree to this because we were discussing it, it was fun and exciting. It was experimental chemistry, descriptive chemistry in those days. How they did things was very exciting and I urge us to think more about that. But with that background, I entered what was then Oregon Agricultural College, a small land grant school. And I found they didn't have a course in chemistry; they had a course in chemical engineering. So I took chemical engineering. That's all.

A.E. Brown: Thank you.

W.J. Bailey: I'd like to follow up on some of your educational experiences. We all are influenced by our early teachers and, you gave one example. Could you follow up and give a few more examples?

M. Harris: Yes, as an undergraduate I took this course and I was particularly influenced by a man by the name of Floyd Rowland who was chairman of the department. He was an organic chemist, I believe, out of Illinois. I don't know what he was doing running a chemical engineering department, because it wasn't much of an engineering department, in modern chemical engineering terms. But we had chemistry courses offered in that department. I was only 17, and when I was a sophomore, he called me in the office one day. I was a bit frightened and he said, "You know, you've just been piling up more courses than anybody takes here, and you've taken most of your courses. If you can get through organic chemistry, you'll be finished your college career here next year." And he said, "I can tell you, I don't know that you can afford to do that, but the University of Washington, (they had an outstanding chemistry department at that time; Dr. Dane and Dr. Powell were really powerhouses in those years) and they have a 12-week course and if you can get through that you'll be finished here next year. You've taken everything that we have. And with that encouragement, I went there. But there was another thing he said, "Now when you get through, what are you planning to do?" And I said, I'm going to get a job. My parents have this little store, which wasn't very effective. But the only job I know is that they hire a couple of analytical chemists at the gas company in Portland. Those were the only chemical jobs around that I knew of. Here I'm reminded of Speed Marvel's experience because I read something about him that sounded very similar. Anyway, my professor said, "I think you ought to go to graduate school." This is absolutely true--I said, "What's a graduate school?" He said, "Well, you take advanced courses, you go on and you learn to do research." This is also very true, I remember saying, "research? We know all there is to know about

Milton Harris as a graduate student at Yale University.

Milton Harris (right, first row) receiving the Wilbur Cross Medal from Yale University.

chemistry. Look at all those books we've got in the library." Anyway, he was persuasive, and he said, "I hope you can go east because there is a lot of activity there and the whole story about research in top universities. One of the former students from Oregon State is here, Tom Hawley, whom you met last night. He was a good friend of mine who went to Yale and so I went along to Yale. It was very difficult, of course, during my first year, coming from a little small western school to Yale, where you were competing with all the Ivy League boys. But after the first year, it went very easy. What more should we talk about?

W.J. Bailey: Well, what were some of the influences you recall at Yale?

M. Harris: Well, Yale had some profound influences on me because I was old enough then, 19 years old, mixing with graduate students, I got a lot of encouragement early, i.e., I met Pete Johnson, one of our classical organic chemists who was doing some synthetic work and he was determined that I was going to work with him. I really wasn't crazy about the synthetic work. Fortunately Yale had this fellowship in biochemistry that I was studying with Professor Vickery. I think the fellowship was for a thousand dollars, and that was an awful lot of money. Most fellowships were for three or four hundred dollars. Tom Hawley had this thousand dollar one before I did. That got me interested in protein chemistry and structure of Protein Fibers too. I found it very interesting and the work was very exciting. We didn't work with silk in those days, that came later. But, it was natural that as a result of that research I became interested in fibers--it just shows on what flimsy threads (I'm not making a pun right here) our lives are run.

W.J. Bailey: Is that really what started you off on textiles?

M. Harris: Yes, that started me off on textiles. My first job was in the Cheney Brothers Mill and they had Tom Hawley working there too. See our careers go back a long way together.

A.E. Brown: Maybe Milton thought that textiles were a high profit industry because of the lucrative Cheney Brothers scholarship.

A.M. Sookne: What happened around the time of World War II? It upset the lives of a lot of people. How did it affect you and the work that you and your group were doing?

A.E. Brown: First I think you ought to tell us how you got from Cheney Brothers to the Bureau of Standards, a couple of us are very interested in that shift, and we just heard a little bit about it yesterday. How did you get from the Cheney Mill to NBS?

Dr. Harris receiving National Bureau of Standards award from Richard Roberts (right).

M. Harris: Well, I think it's very interesting. You know, Pete Johnson was very happy to have the Cheney Fellowship money for his students at Yale. So he was very happy to have me at Cheney to show them that Yale graduates were good employees for them,too. I got a telephone call from Pete one day and he asked if I could come to see him at New Haven. I did and he said that he'd been on an Advisory Committee with a textile industry group. They were going to start some textile research in a group at the Bureau of Standards. There was a system at NBS in those days of trying to get industry, especially those who were not doing very much research, interested in supportive research. The industry foundation paid the salary of the researcher and the Bureau of Standards supplied the laboratory facilities. And the work had to be all basic research. That turned out to be a very, very exciting opportunity. The foundation had a lot of other fellowships at various places and they decided to put them all together at the Bureau of Standards. Being the only "man on the street" at the time, I was made Director of that research group. There we did a lot of very exciting work on cellulose, silk, and wool, which is a fabulously exciting molecule, with a 3-dimensional molecular structure. One of the nice things that was going on in Washington about the same time was the work of a young fellow, Vincent duVigneaud, who Speed Marvel knew well. duVigneaud was just beginning a small Biochemistry Department at GWU and he was concentrating his efforts on insulin. And we were both shocked to find out how similar the chemical structures of wool and insulin were. Both were 3-dimensional polymers held together by the cystine disulfide linkage. In those days we had a little private symposium that we held in the basement of our house. My wife, Carolyn, supplied the beer and pretzels, and about a dozen people from NIH, NBS and the universities met about once a month and discussed different aspects of our work. This went on for several years and was very productive. I don't have to tell you out of that NBS Research Group came about 200 papers. Of course, we got into cellulose, in a big way too; it was a very exciting career. But that brought us up to just about the war, which had a profound effect on us. It changed my whole philosophy. We were immediately pulled into advising the Army Quartermaster about textiles and the Bureau of Standards had a myriad of problems. We had a lot of very early successes and this really impressed me more than anything in terms of my own future industrial career. We made practical use of our research. We were really, I'm not going to be modest, very good and solved many industrial problems.

D. Baker: Why did you start abstracting articles for Chemical Abstracts in 1932?

M. Harris: Well, a very interesting thing. There wasn't very much given in textile courses. I had to educate myself and this abstracting of literature was a good way to learn. I wanted to learn more about textiles. If you want to learn about something, abstract papers from that field. So I was abstracting, mostly textile research articles, and then I was made editor of the whole section 25 on Dyes and Textiles. There was a lot of surface chemistry, too, and I dove into it then. This is a wonderful example, in retrospect, of how all these extraneous things really steer your life. You don't really plan it, it just happens. Somebody once said, getting on in the world is half luck and half ability. And I'm a great proponent of that philosophy because there has been just an enormous amount of luck and good fortune in everything that I've done.

A.M. Sookne: How do you see some of the other side of that statement?

M. Harris: Well, I met many people such as Charlie Overberger and Speed here. I had a brief experience of being sort of a professor, because for several years I gave a course at Catholic University in polymer chemistry. Now just before, let's say 4 or 5 years before that, I got to understand polymer chemistry and I decided the way to look at fibers is from the polymer point of view. There was no such thing as a polymer chemist, or polymer courses, when I was at the University. So I learned polymer work a little later in life. I read a great deal, went through Meyer and Mark's books, and all of the polymer books that were around. But the best way to learn a subject is to teach it. I figure in those days that I spent between 10 to 15 hours in preparation for every 2-hour lecture I gave. And I had some great students, whom I encouraged to go into polymer research. Leo Wall was one of my students, and Dan Frishman, who later came to work at Harris Laboratories. He was in that course, and he's been a very creative and productive scientist in polymers, fibers, furs, and now plastics. And there were many others, but time doesn't permit naming them all.

A. E. Brown: Well it's interesting, Milt, to hear that background. We're all aware, I think, of what Helmut Zahn said so very well about you yesterday, and I would agree. From the early times I was with you I saw that you knew how to pull a lot of things that seem unrelated together, see the linkages, and make something happen. For example, how many people would see the connection between insulin and human hair and wool? Very few, unless you read the literature in all of these fields. You and Vince du Vigneaud saw that relationship. So you're right. To teach a course is a great way to learn. Also another great way to learn, as Helmut showed us in his lecture yesterday, is to accept an invitation to give a talk. I find that the most useful way I know to make me study the literature. And Helmut said yesterday he read all of Milton's papers to prepare his excellent talk. I think Helmut deserves a round of applause for that. I had to read all of Milton's publications on wool and all of J. B. Speakman's articles on wool and hair as an expert witness in a Gillette patent litegation suit involving the chemistry of hair waving. It took me three months to read every paper and I was a knowledgeable man on Milt's wool work in 1952. Milt, you're really getting to the end of WWII. As I mentioned at the Banquet last night, I ran into Milton during the last year of WWII when we really had some rough problems in the South Pacific. Electronic equipment was shorting and the Navy couldn't get messages across the Pacific. Gordon Kline was at the Bureau with Milt, working to solve many of these problems.

M. Harris: Let me get back to WWII. Early in WWII, a man by the name of George Dorio, a Frenchman, who was at the Harvard Business School, came down to Washington as a Colonel to run research in the Quartermaster Corps. And if any of you knew him, he was one of the great characters of all time. One Sunday morning at 8 am he called me and said "I'm Colonel Dorio. Everybody said I must call you. We have a very critical problem, and I'd like to see you." And I said, "When?" He said, "NOW." This is the way he operated. I immediately got in my car and went down to the Pentagon. When I came in he said, "We've got some very critical problems." He was very brusque. He opened a cardboard box and in it was typical wool, or cotton underwear. He was a very dramatic person. He put his hand in it, pulled out this garment and it just fell apart. He says, "What happened?" I laughed and I said, "Colonel Dorio, let us get one thing clear. I'm a chemist, not a fortune teller. Tell me what it is." He said, "I can't. It's for chemical warfare. It's top secret." I said, "I think if you can tell me what's been done here, I might be able to guess or find out pretty quickly what's happened. We've worked on degredation of cellulose from a lot of points of view: photochemical, bacteriological, and everything else. And you know, he immediately got on the phone to some four star general, and I'm just sitting there, and he said it's critical that I get this man cleared for top secret. And he said, "Now," and he says, "We can do it tomorrow morning." And by golly, Monday morning he did. This underwear was being treated with some chemicals to protect against gas warfare. Hydrochloric acid had been liberated which degraded the fabric. At the NBS we had been doing a lot of work, Henry Rutherford and some of the boys, just basic work on cellulose. So in about a week we had an answer for the General. By golly, we became heroes. And then we got together some of the DuPont people who were also put on the problem. And then there was this rotting of cotton fabrics that you referred to.

A.E. Brown: Yes?

M. Harris: Sand bags and tents.

A.E. Brown: Let me take you back to that, Milt, because I think it's very important. They gathered a number of academics at GW, a botanist, a microbiologist, and organic chemists.

M. Harris: Glen Greathouse, too.

A.E. Brown: Glen Greathouse, who was a microbiologist from the University of Maryland, and these generals would come around and say, "We've got this urgent problem." Now I was working for Gordon Kline then not for Milton but Charlie Heimsch and Lee Shanor were working for you.

M. Harris: We were all chairmen of committees.

A.E. Brown: Milt and Gordon would show up about once a week.

M. Harris: Oh, every day, Sundays, Mondays, Tuesdays, the way people worked in those days was incredible.

A.E. Brown: Well, Milton would come around for bull sessions at the end of the day to discuss many urgent problems we had. Because as I look back, like Milton, I was going to go back to the University and be a professor too, but I became such a problem solver under the urgency of war and enjoyed that so much that I never got back to academia. Solving problems under pressure because we had to have answers fast was so exciting.

M. Harris: We learned much. I can cite a dozen things that we solved in a matter of weeks or months and the reason was that our group was so effective. We understood the fundamentals of wool structure and its chemical reactions. And the message that I learned was that if you want to do great applied research, you should first get to know the basics. If you understand the stuff, you've got a much better chance of solving most of these things. And I abhor, I know you've heard lectures from me on the subject, the words "basic" research and "applied" research. I want to talk about knowledge that we need. In every university I'm involved with, that's all I keep stressing. I think the words are counter-productive; you need them both, and if you want to get a society that works, you want the best basic research and the best university work and the best people who know how to use it.

A.M. Sookne: As I recall, you said that you were starting an independent laboratory because you saw a need to do something. Could you tell us about that?

M. Harris: Well, I wasn't so clever, everybody thinks it was very clever, and some made a great deal about it. The truth is that WWII finished on schedule. We all knew very early in January that it was going to end, and that the Textile Foundation

Milton Harris

was going to be moved to Princeton. I wanted to stay in Washington. Princeton was too tough an academic atmosphere for me. No, that wasn't really involved. A lot of things were happening at the time. Salaries had been frozen during WWII. Most people forgot that. I was offered the job of head of the chemistry department at a pretty good school. I asked them how they would keep this faculty together on the starvation wages that they paid in those days. They were incredibly bad. I was leaning that way. But one of these textile companies tried to get me involved in one way or another. Then I was approached by The Gillette Company. They knew they were cutting "wool" fibers off of people's faces. What they were really afraid of is that because of research we had done on softening wool, and some other people were going to invent solvents, (this is over simplifying) and they would be out of the razor and blade business. And I assured them not to worry too much because while the hair fibers were keratin, so were the epidermal tissues and skin and that if you were going to dissolve hair with a chemical you might dissolve the other. But I suggested a study to them. I even sent Chase Jones, who worked in the Department of Agriculture, up there. They were looking for a protein chemist. Their total technical staff consisted of a chemist, with a bachelor's degree from MIT, a physicist named Lou Young, Vice President of manufacturing, and then a few technicians. That was the whole research lab of the company and it was a 30 or 40 million dollar a year business in those days. But then they said whatever you do would you work with us, either as a consultant or per diem, or we'll build you a lab. That started me thinking. Then the American Enka Company, which was owned by the Alien Property Custodian, approached me. I had been down there and given a few lectures on cellulose. They again asked if I would be interested in working with them. So I thought I'd try this laboratory of my own, scared to death. Arnold was one of our boys, and I said look, I'm going to try this. If you fellows want to do it, we'll try it. I was modestly well known in the textile industry by then, and if it doesn't work out, we can always get jobs.

A.E. Brown: Milton, you said one of your first dilemmas was concerned with joining Gillette. Would you like to tell us about that?

M. Harris: Well, it was a very complex thing. The Gillette Company was growing very rapidly in those days and I got to know the top management very well. As they kept expanding, the Chairman of the Board, Carl Gilbert, used to call me and talk about their plans. Then, they acquired the Toni Company, which was all based on polymer chemistry; that is, breaking of the cross linkages in hair to soften and curl it. A wonderful example, real basic research being turned into an applied problem. And then they bought the Papermate Pen Company, that is based on ink chemistry, pigments and surface chemistry. In each case, the Chairman asked me to go out and spend a week in each place, and come back with a research program, because he was a great believer in science and technology. He said, "we don't want anyone catching us asleep technically." As he said in those days, "we paid 20 million dollars for the Toni Company and when somebody sees that we bought a company for 20 million dollars, they're all going to get into that business next week, and I want to be ahead of them." So I helped organize those laboratories and hired people to work for them. And then about 8 years later, in 1955, he called me in and said "Look." (He was just the President then. He was going to move up to Chairman.) He said, "Look, I'm changing jobs. This is kind of big, and you built this whole thing from the outside, now run it." I said, "Carl, this is great! I'm flattered, but the Harris Laboratories are bubbling over. We were having a lot of scientific, technological and practical successes, truly great. Well, he wouldn't let go for about six months and then he convinced me. He said, "We're going to take over Harris Labs." By this time they were the biggest client in our Laboratories. This was 1955, and so I agreed to join the Company as Vice President for Research. They did one very interesting thing. He said to me, "Nobody's going to tell you what to do at Harris Laboratories, you study it for six months; don't make any rapid decisions. You come in and tell us what to do." I offered them a plan that we followed for a long time, which was that half of the work at the Labs would be devoted to problems of the Company as Central Corporate Research, which was typical of what companies were doing at that time, and the other half would be contract work for others. Projects had to have some chemistry or something where we'd be learning things all the time.

A.E. Brown: You were talking about where the contracts came from?

M. Harris: A few came from government and an awful lot from other non-competitive companies. I don't remember them all, Al, you probably do. I remember Dow was one of them. At times we did work with Carbide, Dupont and many others.

A.E. Brown: One of the problems that Milt worried about at the time was that one. He said, "Look, if Gillette buys us out, no other company might want to work for us, which is possible." So he said, "Ok, let's set a policy. We won't work for any competitor no matter what, and if we have a question about whether it's a conflict of interest, we'll send it back up to Gillette people in Boston." We knew everybody at Gillette very well and that helped a lot. So we had contracts with DuPont, Carbide, Dow, Allied, and lots of small companies. For example, Lou Mizell would be working with some small wool company, Dan Frishman would be working with a fur company, whereas Arnold would be working with American Enka and Tony Schwartz would be working with

Milton Harris, while Chairman of the Board of the ACS, conferring with Robert W. Cairns, then ACS President.

Dr. Harris (right), then Chairman of the Board of the ACS, with Dale Baker (left) and Fred Tate (center).

Dow, two large companies. We had no problem with contracts because those were great years for research growth. Nobody was competing with us as much as today, thanks to this fellow.

M. Harris: It's easy to sell something you really believe in. It's as simple as that.

E. A. Calnan: You then came to England and set up a Gillette Laboratory. We know you've always had lots of English friends; you still do. Would you like to tell us about that Milton?

M. Harris: Well, that wasn't very complicated. Gillette was a world-wide company and around 40% of our income was generated in Europe, which also furnished products to the Near East in those days. That developed into a problem of transferring technology. Most of the research was done in America, but some of it was done in England. Gillette had some very good engineering over there. But this business of transferring technology from the U.S. to England--you know Churchill once said the U.S. and G.B. are like two sister countries divided by a common language. The difficulty of transferring know-how and everthing was very real. And the best way, the only way you could to it, was with technical groups. And I decided to set up a laboratory in England as a way of transferring technology. Of course, later, some of our outstanding developments of today came from their laboratories. This concept of using two blades, and that's not a gimmick, everybody thinks that it's all advertising, but that's a real breakthrough started in England. But we started those laboratories and Carolyn and I lived over there for almost a year once, while the Labs were being finished. That's when I found Ted Calnan, who was a young metallurgist and he became our first Director of Research.

B.R. Stanerson: Well, this is all very interesting, Milton. But it was about this stage, when you started thinking about retirement, that a completely new set of circumstances came about and the ACS was a major recipient of this. I was wondering if you might want to say how you made this transfer?

M. Harris: It was very accidental. Actually it started a little bit before my retirement. I was told that my name had been brought up for Chairman of the ACS Board of Directors and one day I was asked if I would accept nomination? I declined, because at that time I was spending almost 30 percent of my time in Europe alone. But I was asked if I would leave my name on the list of possible candidates for the ACS Board and I said, "Ok" and went on to England for one of my usual trips. There was an election soon thereafter and I was elected to the Board, and so I was invited into the first meeting, and shortly thereafter

Art Cope, who was Chairman of the Board, died suddenly. And the Bylaws said that the President, who was Bill Sparks, a great friend of all of ours, great polymer man, should call the Board together and have an election. I was a freshman and wasn't really interested and all of a sudden I found that I was Chairman of the Board. I was elected with the thought that 5 more years was a little too long. I think we've got that changed now, but Stan was then Executive Secretary of the Society and I really got involved. But it was one of the really pleasant parts of my life, and I enjoyed it very, very much.

B.R. Stanerson: It came at a very opportune time because I was a relatively new Executive Secretary and I found it so much easier to have somebody close at hand in Washington to work with. The previous Chairman of the Board had been out of town and, of course, was always available by phone, but in town mainly for the Board meetings. And to have somebody when you're new in a job, to have somebody that you can just turn to day-by-day, was really a great asset to me and I appreciated it.

M. Harris: It was a very exciting period for me. Well, I didn't really know much about the workings of the ACS and within a month after I was named Chairman, I became Budget Chairman; and I remember very well one of the things that stared me in the face was a 2 million dollar deficit up in a place called Columbus.

D. Baker: $1.5 million, I'm sorry, a million five. And you sat me down in the Hilton Hotel in Chicago and said, "Dale, you've got to change things."

M. Harris: Well, I decided, that was one of the problems. I found myself up to here with a budget in those days of maybe $50 million. It's about $75 or 80 million now isn't it. Was it more then?

B.R. Stanerson: No, maybe $30 million.

M. Harris: Well, ok, maybe 30, something like that. At any rate, so I said to Dale, "Well, it'll give me a good chance to learn about the business that we're running here," and let me tell you, I learned more about Chem Abstracts. I want to say it's a super organization, an abstracting organization unlike any in the world. And I'm happy to say Dale got to know more about Chem Abstracts than he ever knew before.

B.R. Stanerson: Well, let me say that we all benefited from Milt's service.

A.E. Brown: Milton's had a pretty darn good crystal ball. I was telling him the other day about a quote which I recently read that surprised me. It's a quote from General Robert E. Lee, who was said to have lamented almost 100 years ago. "Even the future isn't what it used to be." Many people are always pessimistic, but this character Milton--I've never seen him pessimistic about the future. It's 1981 now, Milt. We've got lots of problems in the country and in the world, and in science, Milt, what do you see in the future?

M. Harris: Well, I see a lot of problems, but there'll always be a lot of problems because there are a lot of people. Problems are nothing unusual. You watch your family grow up; you have problems; there's nothing new about any of these things. What we forget is that the problem business is par for the course. The world is getting to be a better place to live in. But today you see troubles instantly when there's terrorism. 15 minutes after it happens, you see it 10 times during the day on TV and in newspapers. If there's a war in Angola, you'll hear about it. But 100 years ago there were wars going on constantly, but nobody knew about them. It's not very long ago, in terms of history, when we all lived behind stockades. We don't live behind stockades now. When you hear about a little mugging, you think that's the end of the world, and it's not so long ago that you couldn't travel much by ship. Until Britain started patroling the seas, that's the first time you really started moving around the world. So you've got to understand that, as long as there are people on the world, there'll be problems. Well, I was told you have crime because you have no education, not enough education. Now 50 percent of our people go to college and you have more crime than ever. So, I don't want to draw any conclusions. And the ups and downs, and you can't keep a balanced economy. The only way you can keep a balanced economy is in a totalitarian state and none of us want that. So you've got to understand that's par for the course, and therefore what we do in the capitalist system works pretty well, give and take. You overproduce, and we all know about that in the synthetic fiber business. Business gets pretty good and the machines start running 24 hours a day. Two years, three years later you've got too much, and so you slow it down. Five or six years ago in Fortune magazine there was an article, about real estate in New York City and how there were millions and millions of square feet empty space. People were already beginning to talk about the demise of New York City--they really were--this is not rhetoric. How can New York survive? You go up to New York today and it's bubbling. I had lunch with a friend up there, just happens to be in the real estate business, he's purely a social friend, just about a month or so ago, a very big man, to to speak. He told me that if he had to deliver 20,000 square feet of space today he couldn't do it. Now if you go up there you'll see the cranes going full blast, with lots of building. Four or five years from now there'll be some vacant space, there will be too much. It will slow down. We in the academic world, we

Milton Harris...

...Captain of the Oregon State University tennis team

...golf champion

...sailor

quadrupled our university populations during the 50's and 60's. I don't know what we did to the graduate schools. Everybody had a graduate school, and all of a sudden we had too many graduate schools. And things are always going to be this way. In our kind of system, it gets out of whack. Today though, these commentators on every television program (and the adversary position has become the popular one) highlight the negative that seems to appeal to prople. Good news doesn't make the news, but I don't think the world's getting to be a worse place. We've got a lot of problems to solve. We've got a crisis of expectation. Nobody wants to give up anything, the universities included. And I'm not saying this in any derogatory manner. I'm very involved with universities, more than a third of my time is just given to a certain university. We had this tremendous growth during the '40s, '50s and '60s, and it was really a golden age. But you can't live on an exponential curve. There's just no way, no brand of economics, could allow you to live on an exponential curve. I had the good fortune, in connection with a talk I was going to make, to visit a couple of dozen laboratories, friends of mine, you know, GE, Carbide, etc. I am convinced, absolutely, not raw optimism, that we are entering a new golden age of research. Think about it; you just watch the things going on in the life sciences and in the electronic field, computers, in the field of mathematics, and the field of astronomy. It's not that people are any smarter, they're good, they get more educated earlier. Freshman today take what I took as a graduate student, if I got the course at all, but what they've got is better tools today, better instruments and a lot of know-how. I am really optimistic about the future; I'm very excited about having the opportunity to watch it for a few more years, and even to monkey in it a little.

A.E. Brown: Well, I want to thank all the members of our panel and our star performer here, Milton, for a very interesting hour. And I want to thank Norm Hollies, one of our old colleagues, for making this hour possible on this morning's program. I think that by pushing it up to before lunch we had a very enthusiastic audience. And, Milt, thank you very much. We'll all be back five years from now to celebrate your 80th!

RECEIVED April 5, 1982.

2

Fabricable High Temperature Stable Polyaromatic Ether Ketones and Polyaromatic Ether Keto Sulfones

C. S. Marvel

University of Arizona, Department of Chemistry, Tuscon, AZ 85721

Some time ago, we determined that polyaromatic ether ketones and polyaromatic ether keto sulfones were very heat stable and oxidatively stable but they melted too low for practical use.[1] We were interested in finding some ways to introduce a cure site into the chain, and thus make them more practical polymers. We tried many different methods of introducing cure sites into the polymers to render them capable of crosslinking to raise the softening points so they'd fit our use. First, we tried to introduce cyano derivatives into the chain by end grouping it with p-cyano benzoyl chloride and by replacing isophthaloyl chloride by

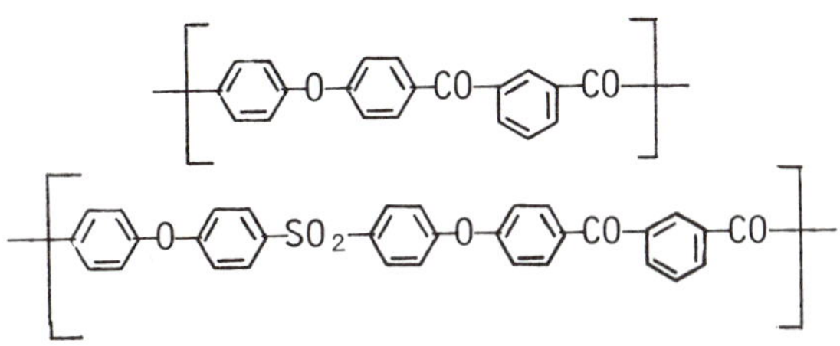

cyano-substituted isophthaloyl chloride, and finally by dicyano-substituted terephthaloyl chloride. We could not get good cyclizations of these cyano groups in any case, and the cured materials were always less oxidatively stable than the starting materials.[2]

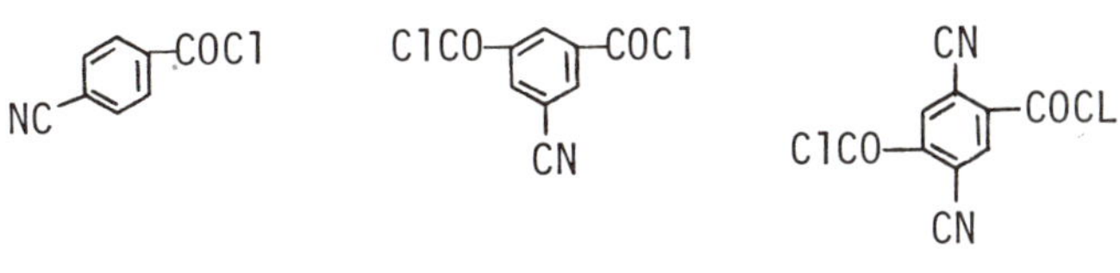

Next, we substituted cyclophane units into the polymer chain and attempted to crosslink by opening the cyclophane units by heat. This worked well except that introducing the benzyl hydrogens in the cyclophane made the products too readily oxidized for our use.[3]

8412-0740-2/82-0021$06.00/0

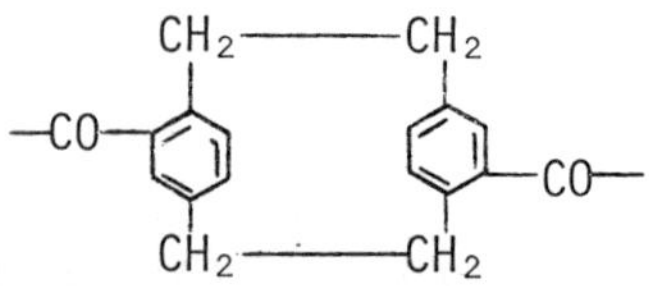

We next attempted to introduce acetylenic groups which we hoped to trimerize to benzenes and thus crosslink our polymers. For this we used p-acetobenzoic acid, which we converted to the acetylenic compound after end grouping the polymers with this

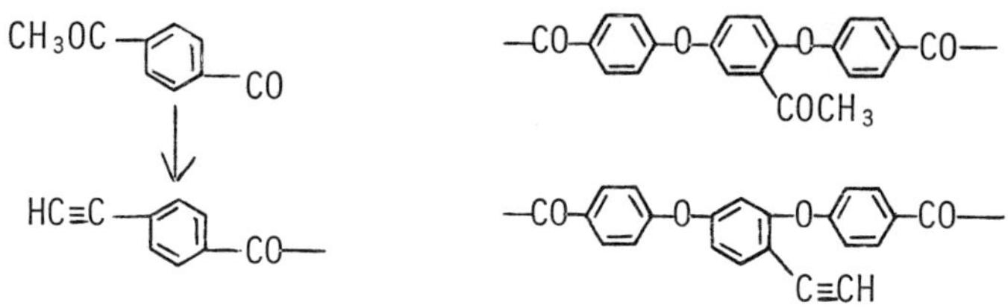

unit. We also made an acetyl derivative of diphenyl ether of hydroquinone which we introduced in place of some of the diphenyl ether and converted the acetyl groups to acetylene groups to get these scattered along the chain. These materials crosslinked, but were less stable to oxidation after processing than before.[4]

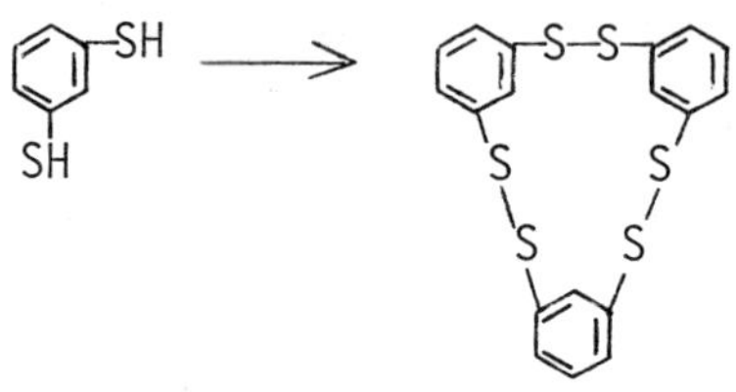

We introduced cyclic disulfide units into the chains and they could be used for crosslinking, but the products which we prepared were too readily oxidized to be useful for high temperature polymers.[5]

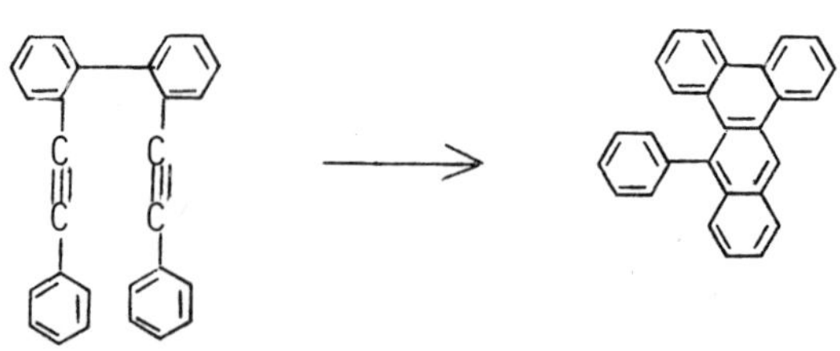

We next attempted to use the Hedberg-Arnold method of intermolecular cyclization of 2,2-diphenylethynyldiphenyl groups, which on heating, are cyclized to phenyl benzanthracene units, and thus raise the melting point of the polymer. This seemed to be too difficult to accomplish to be practical, although it did give good polymers which did cure.[6] We introduced 1,3-butadiene units into

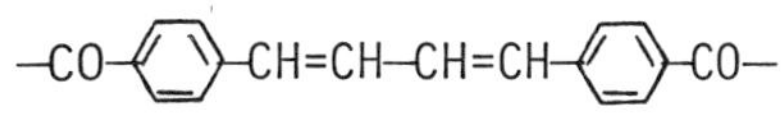

the chain and hoped to use them for Diels-Alder crosslinks, but the intractibility of the polymers defeated our purpose.[7]

We next tried the method of Stille and co-workers,[8] who used biphenylene units to cure their polyquinoline derivatives by preparing polyether ketones with 2,6-biphenylene dicarbonyl groups[9]

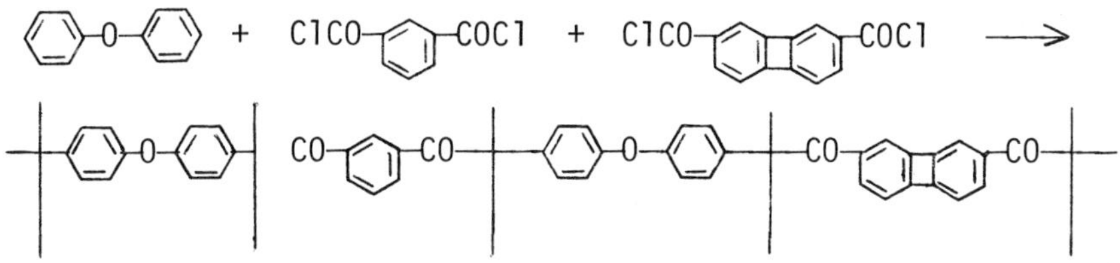

and 2,7-biphenylene dicarbonyl groups,[10] and these will be discussed later. Of all of these, the biphenylene groups seem to be the most promising.

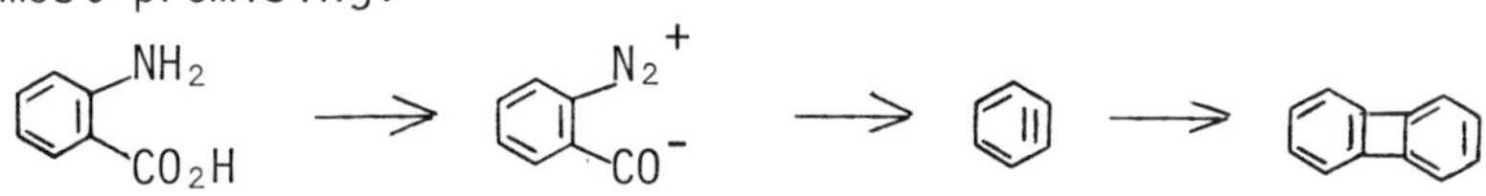

We prepared our biphenylene dicarboxylic acid from anthranilic acid by a diazotization, and then introduced the acetyl groups into the 2,6 positions and oxidized them with sodium hypochlorite to obtain the 2,6-dicarboxylic acid which was converted to the acid chloride.[9]

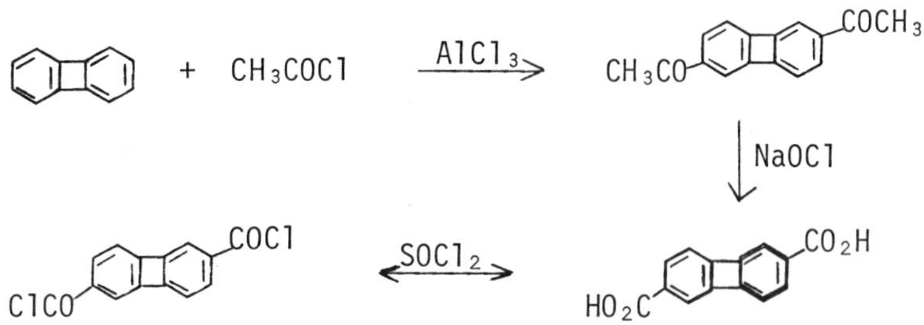

The polyaromatic ether ketones were made by Friedel-Crafts reaction using diphenylether, isophthaloyl chloride, and 5 mole % of the isophthaloyl chloride was replaced by 2,6-biphenylene-dicarbonyl chloride. In a typical run, we used 25 mmole of diphenyl ether, 23.75 mmole of isophthaloyl chloride, and 1.25 mmole of biphenylene diacid chloride in dichloroethane solution with excess of aluminum chloride. The reaction was carried out at room temperature for about 20 hrs.

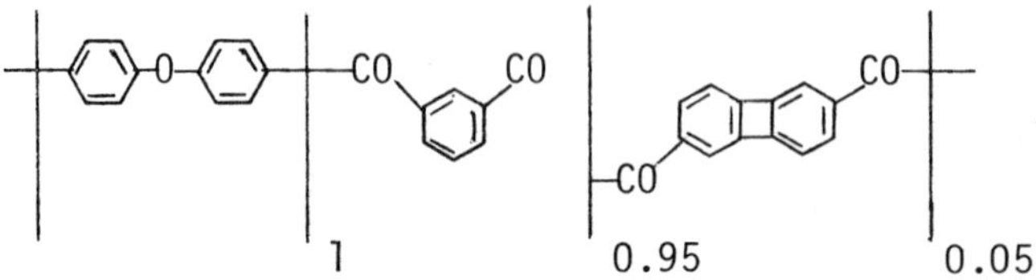

The polymer began to separate at about 20 minutes and the reaction was apparently complete in a relatively short period of time. However, we allowed it a longer period of time to be sure that we'd get a good viscosity in our polymer. Yields were about 94%, and the polymeric material was a light yellow color. It was

> Yield 94%; softening point 240°-270°C; η_{inh} 0.53 in H_2SO_4
> Curing temperature 300°-250°C for 3 hr - insoluble in sulfuric acid. No melting below 460°C.

only soluble in sulfuric acid, the softening point was in the 240-270° range, and the inherent viscosity in sulfuric acid was about 0.53. When this polymer was heated to about 300°C in an inert atmosphere it became completely insoluble and infusable. It has a slight Vicat softening point at about 120-140° but the product did not melt below 460°. When heated in circulating air at 300°C for three days, it lost less than 1% of its weight and further heating to 350° for an additional 24 hrs caused a further weight loss of about 1%. A sample of this polymer was converted to an I-beam by molding at 300°C and 20,000 psi. This beam showed a tensile strength of nearly 12,000 psi and an elongation of about 5.7%.

In some later runs we carried out the first polymerization reaction at about 0° at the start and allowed the mixture to warm up to room temperature after a while. The yields were essentially quantitative. The inherent viscosity of that polymer in sulfuric acid was about 0.8. If the temperature of the reaction is allowed to go too high, some side reactions occur and the product is not too easily processable.[11]

For our first runs we selected 5 mole% of the biphenylene compound as a pure guess, and later we made materials with more and with less biphenylene units in the polymer.

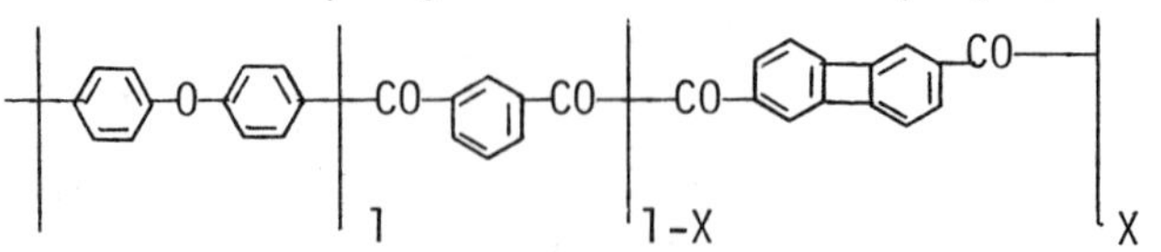

X = 0.025; 0.05 and 0.10
η_{inh} as before. Melting cure unchanged.

For these later runs,[11] we started at 0° and then allowed the reaction to rise to room temperature for some time, and maintained it for about 24 hrs at room temperature. We got some of the products with a viscosity of .99. They varied slightly in the softening point from 205-275°C, and occasionally a side reaction occurred and the material didn't soften at all. Dr. Donald White of our Chemical Engineering Department and Dr. I. D. Steg of the Weizmann Institute,[12] working in our Chemical Engineering lab, made laminates with glass fibers from some of these 5 mole % products, and found that they were extremely tough and stable at high temperatures. They thought they were the best high temperature laminates they had seen, particularly when this keto ether was mixed with Phillips Petroleum polyphenylene sulfide. There was a great improvement in polyphenylene sulfide so that it was not brittle in laminate form. They were much interested in this product and promised to test our samples with lower and higher biphenylene content, but so far their results have not come through to us.

Little difference in properties
for both polymers.

We also prepared a sample of the keto ether using 2,7-biphenylene diacid chloride, which melted in about the same range as the 2,6 derivative and had about the same inherent viscosity.[10] It also cured at about 300° and seemed to be approximately equal in general to the product made from the 2,6-diacid chloride. We obtained the 2,7-biphenyl chloride by the Ullmann reaction on 3,3'-diiodo-4,4'-diphenic acid ester.

Another polymer has been made using p,p-diphenoxydiphenyl sulfone in place of diphenyl ether.[9] The general conditions of preparation and purification of the polymer was as before. This polymer had a softening point of about 255-265°C and an inherent

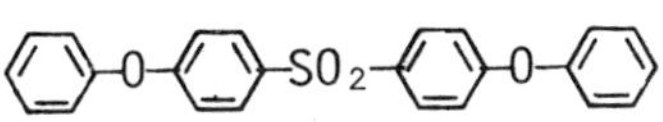

Ketone soluble in sulfuric acid
Sulfone soluble in DMAC
Clear plastic at 300°C.

viscosity of about 0.2. The polymer was soluble in dimethylacetamide as well as sulfuric acid. At equal viscosities it seemed to be less strong than the diphenylether ketone. An I-beam prepared from this had a lower tensile (750 psi) and less elongation (1.7%) than the one from the ether ketone. We do not have an evaluation of this material yet as a laminate.

For both of these products the chief drawback to actual use is the fact that they must be cured at 300-350°C in an inert atmosphere, and the engineers view this as an obstacle which make them unlikely for practical use. Since we are looking at potential use in airplane wing structures, they felt that it was impossible to achieve a cure on a piece of equipment that large in an inert atmosphere at 300° for 3 hrs in a practical manner. Consequently, we made some attempts to find other ways to cure this product. We found that if we mixed rhodium or palladium complexes into the polymer before heating it would lower the curing temperature and shorten the time, but the resultant products were more easily oxidized in air and thus defeated our purpose for using materials as heat stable laminates. We are looking at other metal complexes to see if we can find one that will lower the curing temperature and at the same time not to render the polymers subject to oxidation.

We are also making 1,8-biphenylene-dicarboxylic acid and 1,5-diphenylene-dicarboxylic acid in the hope that the steric effect of these polymers may be such as to make the four-membered ring more easily cleaved, but that remains to be done.[11]

We have two other interesting polyaromatic ether ketones and polyaromatic ether sulfones in which we replace some of the iso-phthalolyl chloride with a meta tolane unit.[13] The ketone is soluble only in sulfuric acid but the sulfone is soluble in di-methylacetamide. These polymers are interesting in that they give

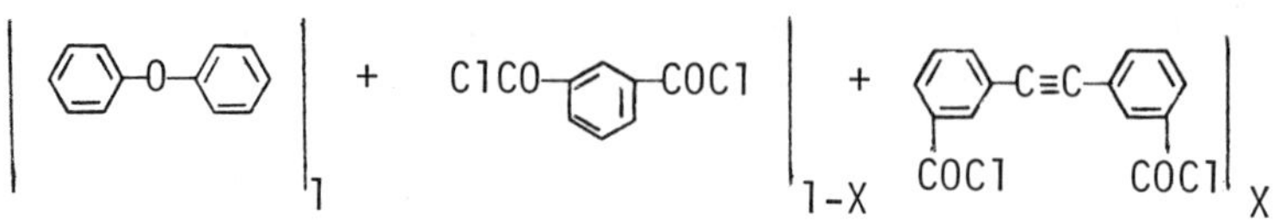

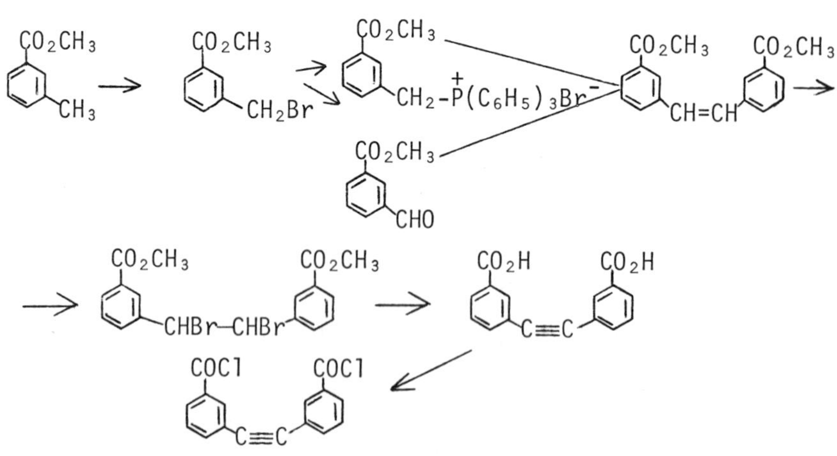

water-clear films which remain clear and colorless to about 300°C.

They have good stability in air also to about 300°C with very little weight loss in circulating air for three days. The films which we obtained are not as strong as we'd like yet, but they are improving with our improved techniques in the preparation of polymers, and we hope we can find some technique for increasing their strength.

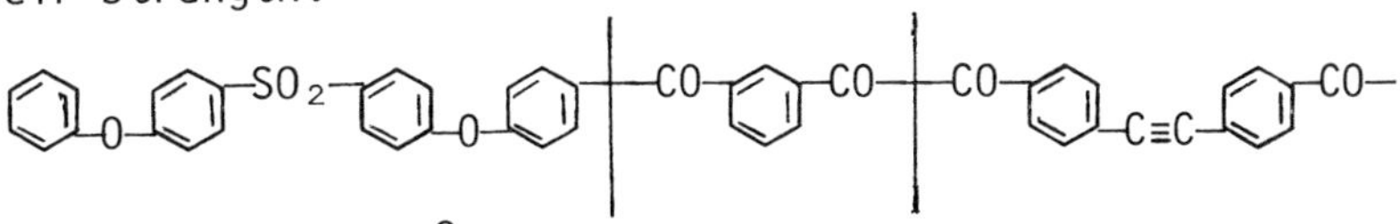

MP over 300°C Soluble DMAC
Clear up to 300°C. Somewhat brittle. Decomposes about 400°C.

Cure with Diels-Alder Reaction

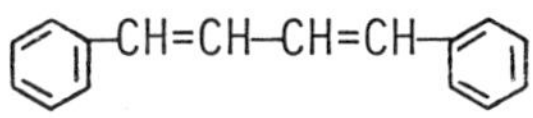

These polymers can be cured by Diels-Alder reaction with 1,4-diphenyl-1,3-butadiene to raise the softening temperature.

We have also used the para tolane derivative[14] with mixtures of isophthaloyl and terephthaloyl chloride, and we have some interesting new polyaromatic ether ketones and polyaromatic ether

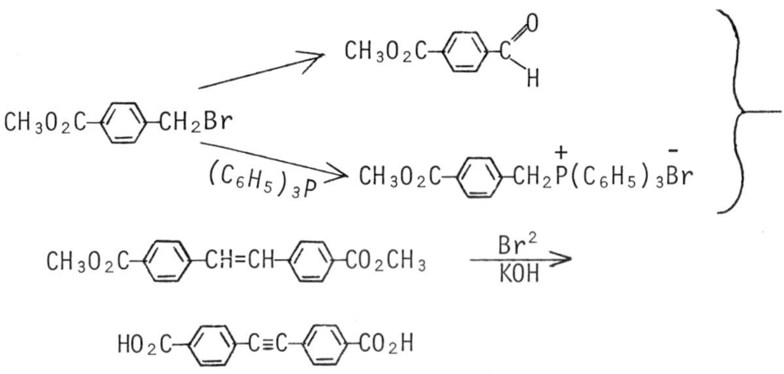

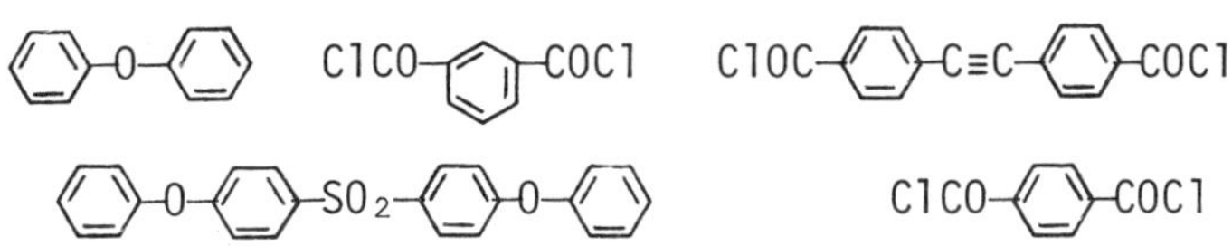

various combinations clear colorless films

keto sulfones in this family, which have softening points in the range of 200-225° and viscosities in the range of 0.3 to 0.5. The sulfones are soluble in a variety of solvents such as DMF, DMAC, and methylene chloride, and give clear films which are stable to 300°. This property was unexpected, but it indicates that these materials may have a use for high temperature clear plastics to replace Lucite or Plexiglas polymers which do not stand up to high temperatures without melting or decomposing.

There is still a considerable amount of work to do before these materials become practical, but they seem to be the best high temperature stable polymers that we or anybody else have been able to prepare and if we can get the curing process down, they should be practical polymers because the only expensive part of the polymer is the biphenylene unit or the acetylene unit.

Acknowledgments

This work has been supported by the Army Research Office and by the Office of Scientific Research of the Air Force.

ARO DAAG29-79-C-0114		AFOSR 77-3112	
R. Swedo	P. Schmutz	S. Lin	B. Lee
V. Sankaran	A. Sutter	G. Kriek	

I am grateful to these people for their help.

Literature Cited

1. Verborgt, J.; Marvel, C. S. J. Polym. Sci. 1973, 11, 261.
2. Sivaramakrishnan, K. P.; Marvel, C. S. J. Polym. Sci. 1974, 12, 651; 1974, 12, 1945; Samyn, C.; Marvel, C. S. J. Polym. Sci. 1975, 13, 701; Bruma, M.; Marvel, C. S. J. Polym. Sci. 1976, 14, 1.
3. Samyn, C.; Marvel, C. S. J. Polym. Sci. 1975, 13, 1095; Sivaramakrishnan, K. P.; Samyn, C.; Westerman, I. J.; Wong, D. T.; Marvel, C. S. J. Polym. Sci. 1975, 13, 1083; Chang, D. M.; Marvel, C. S. J. Polym. Sci. 1975, 13. 2507.
4. Samyn, C.; Marvel, C. S. J. Polym. Sci. 1975, 13, 1095; Kellman, R.; Marvel, C. S. J. Polym. Sci. 1976, 14, 2033; Huang, F.; Marvel, C. S. J. Polym. Sci. 1976, 14, 2785.
5. Wong, D. T.; Marvel, C. S. J. Polym. Sci. 1976, 14, 1637.
6. Banihashemi, A.; Marvel, C. S. J. Polym. Sci. 1977, 15, 2653; 1977, 15, 2667; Frentzel, R. L.; Marvel, C. S. J. Polym. Sci. 1979, 17, 1073; Somers, A.; Marvel, C. S. J. Polym. Sci, 1980, 18, 1511.
7. Sankaran, V.; Marvel, C. S. J. Polym. Sci. 1979, 17, 3949.
8. Garapon, J.; Beever, W. H.; Stille, J. K. Polymer Preprints, Div. of Polymer Chem. Inc., ACS 1977, 18, 1138; Garapon, J.; Stille, J. K. Macromolecules 1977, 10, 623.
9. Swedo, R. J.; Marvel, C. S. J. Polym. Sci. 1977, 15, 683; 1979, 17, 2815.
10. Hanson, M. P.; Marvel, C. S. J. Polym. Sci. 1978, 16, 653.
11. Sutter, A.; Schmutz, P.; Marvel, C. S. Unpublished results.
12. White, D.; Steg, I. D. Unpublished results.
13. Lin, S.; Marvel, C. S. Unpublished results.
14. Lee, B.; Marvel, C. S. Unpublished results.

RECEIVED February 8, 1982.

3

Contributions of Milton Harris to the Wool Textile Science of Today

Helmut Zahn

Deutsches Wollforschungsinstitut an der Technischen Hochschule Aachen, d 5100 Aachen, Federal Republic of Germany

The German Wool Research Institute Aachen takes it as a compliment to have been asked to make a contribution to Milton Harris' Symposium in honour of his 75th birthday.

There are many ways in which appraisal of the work of Milton Harris, a pioneer in the domain of fiber, textile and polymer science, can be attempted. The author has decided to approach his task by studying Milton Harris' publications to find out what effect they still have on current wool research.

The answer is positive: Milton Harris is still 'in', so much so, that citing him is often forgotten because his lines of thinking have set standards of knowledge commonly accepted by scientists in wool science and technology all over the world.

The author was fortunate to be able to revive his memory on many detailed points which he had himself forgotten and to find solutions for his current research problems in Milton's publications issued decades ago.

It was Milton Harris himself who said that he wanted it made clear 'that the examination of present and future needs must never be considered a reflection on past accomplishments'. Nevertheless it is a fact that present wool textile research has a lot to learn from work accomplished by Milton Harris long ago. To prove this was one of the main reasons for the following contribution. Since it is impossible in view of the limiting factors of space and time to give a sound overall view of Milton's research, the author has picked out 13 topics, starting with structural research and ending with clothing physiology, which prove the exemplary relevance that the Harris school still has in wool research.

8412-0740-2/82-0029$08.25/0

Bilateral Structure of the Cortex of Merino Wool Fibers

There are two major types of cortical cells within the wool fiber and in many wool fibers the whole cortex is transversely segmented (M. Horio and T. Kondo, (1)). This so-called bilateral structure of crimped wool fibers can be seen as early as 1941 in a paper by Ch. W. Hock, R.C. Ramsay and M. Harris (2) as shown schematically by Figure 1. Reduced and methylated fibers were treated with pepsin for 10 days. The photograph of cross-sections demonstrated preferential digestion by pepsin of one cortical segment. The orthocortical cell is generally more susceptible to chemical attack than the paracortical cell. A similar preferential dissolution of the orthocortex has been shown by E.H. Mercer (3) on fibers which were reduced and ethylated and then digested with pepsin.

Differentiation between Mohair and Wool

Another example of a forgotten achievement of the Harris team is the differentiation between mohair and wool. P. Kusch and W. Arns (4) have recently described a simple method for unequivocally assigning single fibers to be fine animal hair e.g. mohair or wool, by measuring the thickness of the scale edge under the scanning electron microscope. This thickness was found to be around 0.4 micron for all goat hairs whereas wool fibers of all qualities checked had scale edge thickness values of at least 0.8 micron (Figure 2). We should have read the paper by L. Fourt (5) who already showed that the angle between the scale surface and the fiber axis is larger for wool than for mohair and is independent of fiber diameter, for each type.

New Cross Links in the Cuticle

It was Walton B. Geiger (6) who was the first to study the chemical composition of the scale material and found it to be more stable towards alkali than wool that had been reduced and alkylated in the same way. This was explained by the presence of crosslinks other than those involving sulfur. Many years later M. Nienhaus and J. Föhles (7) showed that the cystine-rich exocuticle contains a surprisingly large amount of such crosslinks, namely the isodipeptide $N_{\varepsilon}(\gamma$-glutamyl) lysine.

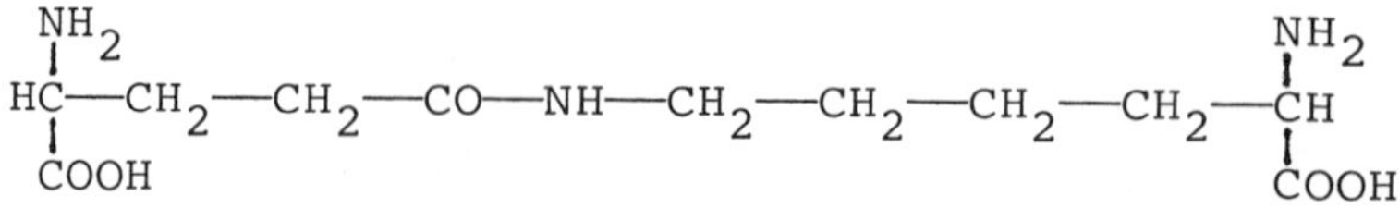

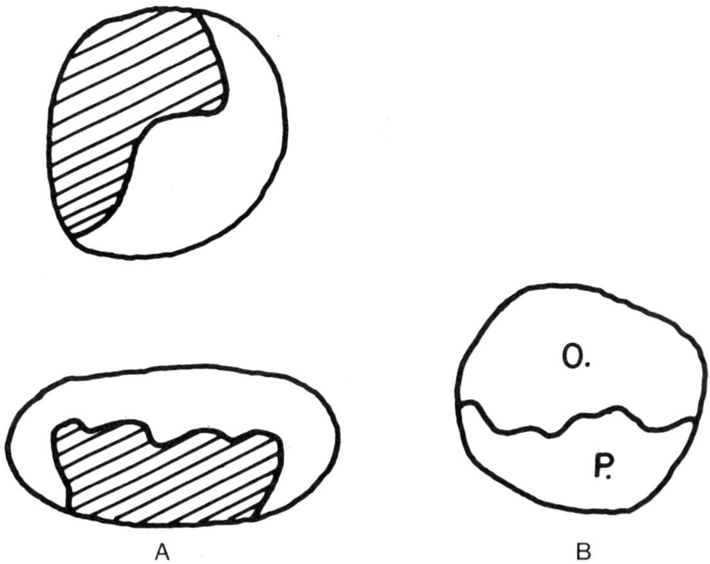

Figure 1. A: Schematic representation of cross-sections of reduced and methylated wool fibers treated with solutions of pepsin for 10 days (2). B: Cross-section of a wool fiber showing the orthocortex (O) and paracortex (P).

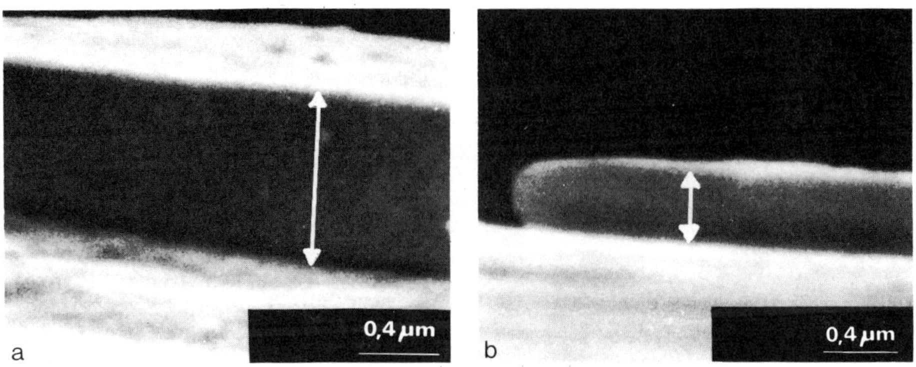

Figure 2. Schematic representation of scanning electron micrographs of the scale edges of a, Buenos Aires wool; and b, Mohair (55).

Lipids and the Cell Membrane Complex of the Cuticle

In the publication by W.B. Geiger quoted above (6) we also found the first indication for a high content of cell membrane lipids in the wool scales. Geiger reported a value of 2.7% for the bound lipid constituent in wool scales. This is much higher than the lipid content of 0.8% in the whole wool fiber (8) but corresponds well with the high content of cell membrane complex in the scale cells of human hair reported by A. Swift and B. Bews (9) considering the fact that the cell membrane complex consists of about 1/3 lipids.

Incorporation of Cystine in Keratin

What is the most influential publication of Harris in the field of protein chemistry of keratins and proteins? In my view it is the paper by W.I. Patterson, W.B. Geiger, L.R. Mizell and M. Harris (10) on the role of cystine in the structure of the fibrous protein, wool. It begins with a concise definition of the primary structure of proteins:

> "Proteins, according to the hypothesis of Hofmeister and Fischer, consist of amino acids joined through peptide linkages to form long chain structures called polypeptides. On the basis of this hypothesis, many of the similarities among proteins should be due to the large number of peptide linkages, while dissimilarities among them should be related to variations in the proportions and arrangement of the constitutent amino acids, to differences in the lengths and arrangement of the polypeptide chains, and to the presence of linkage other than peptide. The present paper is concerned with the relationship of one such linkage, the disulfide linkage of the amino acid cystine, to the properties of the fibrous protein, wool.
>
> Although cystine contains two amino and two carboxyl groups, it appears that all of these groups are involved in peptide links in most, if not all, of the cystine residues".

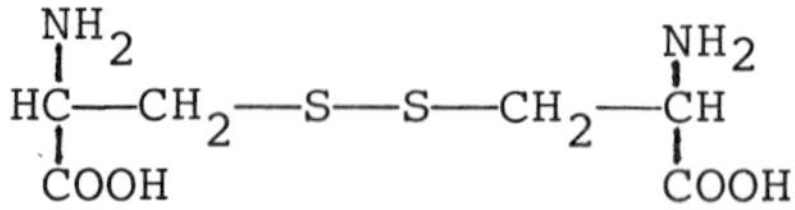

Cystine

The classical studies by J.B. Speakman had already proven that a number of the important physical and chemical properties of wool are closely associated with the presence of the disulfide linkaged between the polypeptide chains of the protein. But it was Harris who was first able to prove this relation unequivocally by finding conditions for specific and reversible reactions at the disulfide bond. Such a specific reaction was reduction with thioglycolic acid over a wide range of pH values. It was established that the fiber does not lose its fibrous structure if the reduction is carried out in acidic or neutral media. The sulfhydryl groups of fibers reduced in this way could be recombined to disulfide bonds. They also could react with alkylhalides to form thioether groups. The breaking of cross-linkages during reduction greatly decreased the 30-percent index of the fibers. Reduction followed by reoxidation resulted in the restoration of nearly the original 30-percent index, an indication of the practically complete reaction of the sulfhydryl groups to rebuild disulfide cross-links. It is possible that this is the first time that a resynthesis of a disulfide protein was accomplished. Biochemists usually quote the resynthesis of ribonuclease from the fully reduced open chain as such a breakthrough (F.H. White (11)).

We found in the classical work of Harris experimental conditions for the preparation of soluble wool keratins in the disulfide form. Such preparations should be of great interest for studying fundamental chemical and physical properties in the keratin proteins. They should be more closely related to the genuine keratin structure than the presently studied S-carboxymethylkerateins, in which the original disulfide linkages are replaced by permanently broken bonds, which in addition are made hydrophilic and ionogenic by S-carboxymethylation. There are many other unsolved problems in the cystine chemistry of keratin, such as the problem of interchain and intrachain disulfide bonds, the SH- SS interchange reaction etc.

Photochemical Decomposition of Cystine

It has long been known that wool slowly decomposes on exposure to sunlight (C.H. Loebner (12)). It was again M. Harris (13) who started fundamental research on this problem. Exposure of wool to a glass-enclosed carbon arc was found to decrease the concentration of cystine. NH_3 and H_2S were evolved during irradiation in the absence of oxygen. The formation of H_2S during irradiation in nitrogen was increased by the presence of moisture, as shown in Figure 3.

Today we know that free radicals produced in solution by ionizing radiation rapidly attack S-S bonds, which are by far the most radiolabile groups in a protein. G. Blankenburg (14) and later Doris Klee (15) found that weathered tips of Buenos Aires wool contain much more lanthionine as well as cysteic acid than the root parts. The fact that one finds products of alkaline as well as oxidative decomposition of cystine in the same sample is only apparently contradictory. The data suggest one possible explanation: that it is the hydrogen sulfide produced from the

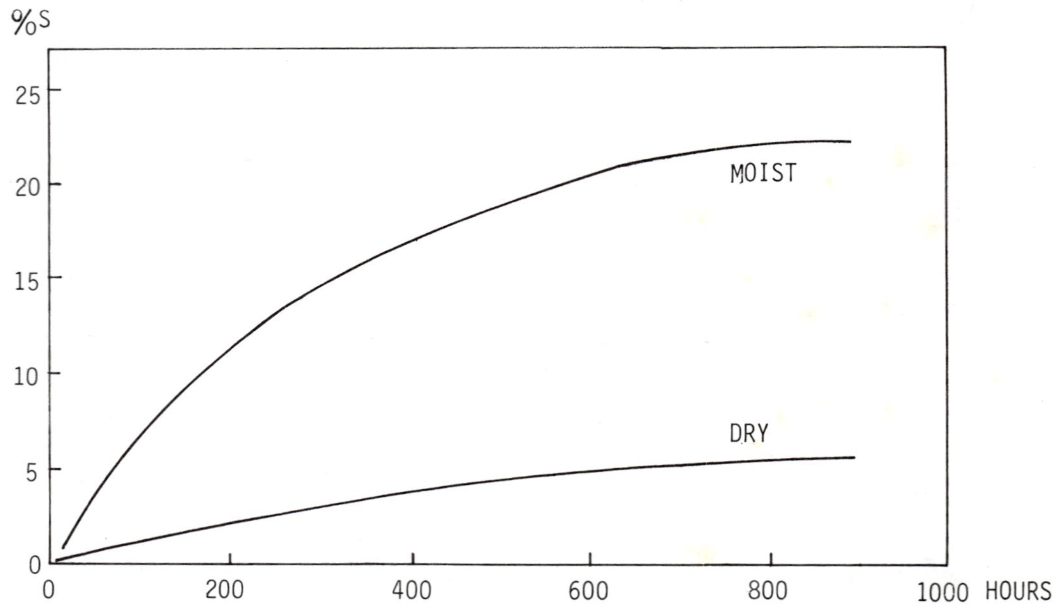

Figure 3. Photochemical reactions of wool. The rate of evolution of hydrogen sulfide from samples of wool exposed in atmospheres of moist and dry nitrogen to radiation of glass-enclosed carbon arc. The ordinate values are percentages of total sulfur evolved as hydrogen sulfide during irradiation (13).

irradiated wool according to M. Harris and A.L. Smith (13) which initiates the lanthionine formation according to Elliott, Asquith and Hobson (16).

Chemical Modification of Wool Cystine

The fundamental publication by W.I. Patterson et al. (10) has already been mentioned. It contains not only the remarkable experiment of a reversible reduction and reoxidation of the disulfide bonds in keratin but also a description of many chemical modification reactions. Reaction of reduced wool with alkyl monohalides such as methyliodide results in permanent rupture of disulfide bonds whereas alkylation with aliphatic dihalides such as methylene iodide or trimethylene iodide introduces hydrocarbon chains between pairs of sulfur atoms of cystine molecules in the fibers. Such wool fibers, in which the alkali-labile disulfide bonds have been replaced by new covalent cross-linkages, possess much lower alkali solubility. "Since the susceptibility of wool to degradation by alkalies is one of the greatest disadvantages, processes that would make it stable toward alkalies should also enhance its durability". This quotation from the fundamental publication can be regarded as a direct invitation to industrial and university research people to enter this field. It was primarily Werner Kirst at Höchst (17) who checked many possible ways of chemical modification of wool along these lines in order to produce wool products with greater stability to alkalies and other chemical agents.

The work of Harris had a great influence on Egon Elöd (18) and myself and I am still proud of having been able to prove the reactions proposed by Harris by direct isolation of S-S-polymethylene cysteines from reduced and alkylated wool (19) in Freudenberg's laboratory (Figure 4).

That the chemistry of alkylation of reduced fibers is still of great interest can be seen in the recent book by John A. McLaren and Brian Milligan (20) who quote Australian work on the influence on the sorption, hygrostress and supercontraction behavior, as well as American work on the stress relaxation.

The new method of replacing disulfide bonds by synthetic cross-links was also applied in order to minimise fibre damage by treatment with sodium dithionite and sodium hydroxymethanesulfinate by carrying out the treatments in the presence of bifunctional reagents, such as 1,2-dibromoethane, formaldehyde or glyoxal (21). The results were interpreted in terms of replacing disulfide crosslinks with more stable crosslinks. The process was claimed to minimise damage during bleaching, vat dyeing, and stripping treatments and to confer resistance to damage by alkali or oxidants (22).

Werner Kirst (23) was induced by Harris' papers to try chemical modification of wool without prior reduction of SS

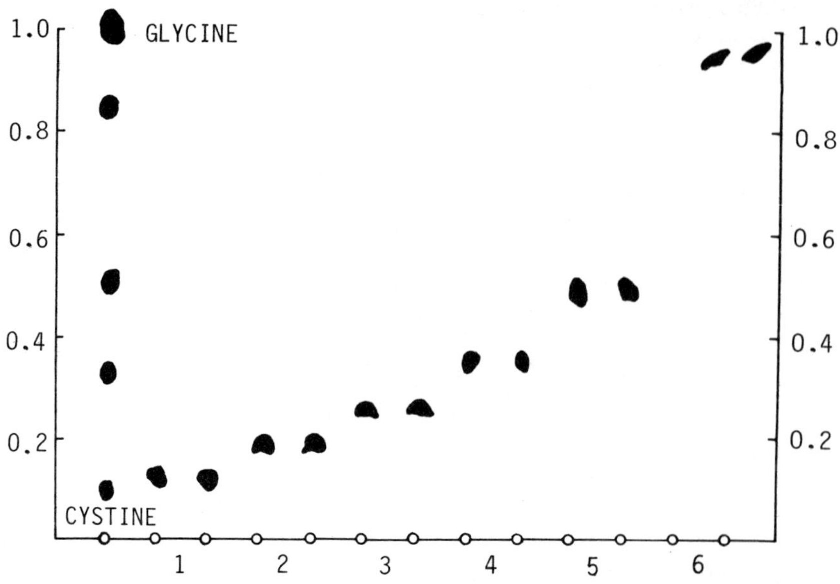

Figure 4. One-dimensional paper chromatograms of cystine, glycine, and S,S'-polymethylene-bis (cysteines) (19). Numbers on abscissa give number of CH_2-groups between S-atoms of cross-linked cysteine residues. Spots on the left are derived from paper chromatographs of reduced and cross-linked wool samples. Spots on the right show the synthetic compound run under identical conditions. Eluent: 1-methylpropanol–formic acid–water (75:15:10).

bonds. A great number of patents and publications again is evidence for the impetus given by the pioneering work of the American research workers.

One might even state that the cross-linking chemistry carried out in Aachen not only in the wool field but also that which finally led to the full chemical synthesis of insulin and to Brandenburg's synthesis of mini-proinsulin (Figure 5) is ultimately a result of the fundamental ideas and methods of the Harris school.

Carbonizing

"Carbonizing, which is employed for the removal of burrs, shives and other vegetable matter from wool, consists in treating wool with dilute sulfuric acid and baking the acid-treated wool to char the vegetable matter so that it can be removed by mechanical action."

With this sentence L.R. Mizell, A. Eleanor Davis and C. Oliva (25) begin their paper "A critical study of wool carbonizing" which turned out to be a landmark for the industry involved in carbonizing, because L. Mizell and associates indicated the critical limits for each factor in raw wool carbonizing and found conditions which completely charred the burrs without causing any measurable strength loss of the wool.

The time elapsed since this publication was well used for practical application of the data for commercial processing. As far as the fundamental chemistry of side-reactions of sulfuric acid with the wool proteins is concerned, it was found by M. Cole. J.C. Fletcher, L.L. Gardner and M.C. Corfield (26) that instead of the N-sulfonation postulated by Harris et al. O-sulfation can be proven by isolation of serinesulfate from carbonized wool (Figure 6). Irene Schmitz was also able to isolate O-serinesulfate from baked wool by enzymatic digestion (27). To sum up, Harris has led the way to a full understanding of the carbonizing process both from the fundamental - except the N-sulfonation - as well from the practical processing angle.

Bleaching

Milton Harris and coworkers were the first to provide quantitative data showing the effect of oxidizing agents such as hydrogen peroxide on wool as a basis for studying all processes in which wool is exposed to oxidation reactions. A. Smith and M. Harris (28) presented data on the effects on wool of varying the concentration, temperature and pH of the hydrogen peroxide solutions and the duration of the treatment. The main point of attack during oxidation is the disulfide group of cystine in the wool. There is a strong correlation between the decrease in the

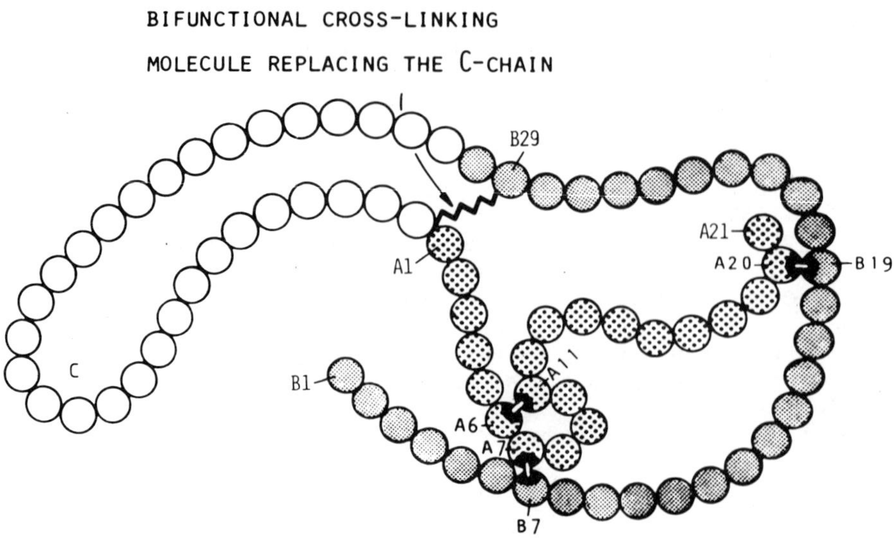

Figure 5. Proinsulin consists of three polypeptide chains, A, B, and C, linked together. The C chain forms a loop connecting the glycine residue at position A1 with the alanine residue at position B30. Two cysteines join the A and B chains by S-S bridges between positions A7 and B7 and between positions A20 and B19. In addition, there is an internal S-S bridge between positions A6 and A11. In proinsulin analogs the polypeptide C chain is replaced by a small bifunctional molecule that cross-links the A and B chains between positions A1 and B29 (24).

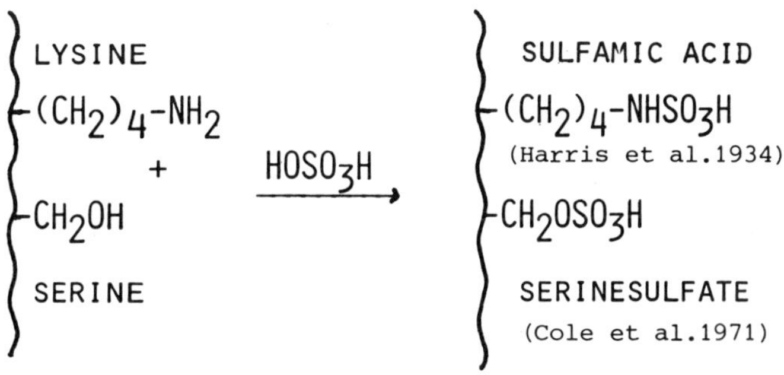

Figure 6. Schematic representation of possible reactions of sulfuric acid with side chains of combined lysine and serine in wool.

alkali-solubility of the bleached wool. Ziegler (29) was able to confirm these results under a wide range of bleaching conditions (Figure 7). Bereck and Schwartz (30) found that this close relationship between the oxidation of cystine and alkali solubility also applies for peroxide bleaching catalysed by heavy metal salts. Critical values for the concentration, temperature, and pH of the hydrogen peroxide solutions were indicated. The alkali solubility test can be held to be one of the oldest and still most often applied chemical tests not only in wool research laboratories but also in the wool industry and even in the wool trade. It was one of the first chemical test methods standardized by the International Wool Textile Organisation.

The work of the Harris school on oxidation of wool and on bleaching with peroxide has not only provided reliable data for wool dyeing and finishing industry but also contains a great wealth of chemical data for fundamental research in wool and protein chemistry. Exact chemical analyses of the total sulfur, nitrogen and cystine content have become mandatory for wool research workers since these publications.

On the other hand there are some discoveries described in the classical work of Harris which are still waiting for further development.

Such problems are:

1. The existence of intermediate oxidation products of cystine, first postulated by Smith and Harris in 1936 (31), was proved by these authors soon after, for instance by reduction with hydrochloric acid-potassium iodide solution (32):

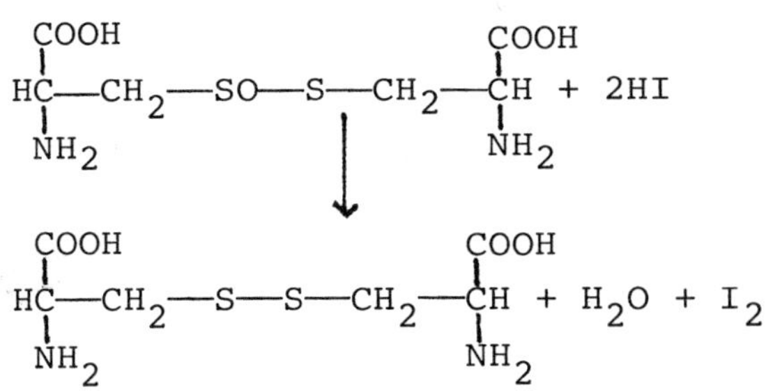

About 30 years later Savige and McLaren (33) were able to isolate the mono- and dioxide of cystine.
Nevertheless we do not fully understand the chemistry (formation, stability) and practical relevance of these compounds.

2. Harris and Brown (34) did describe a direct correlation between the number of intact disulfide bonds and the wet strength of fibers. According to Smith and Harris (28), on the other hand, the wet tensile strength of wool yarn is

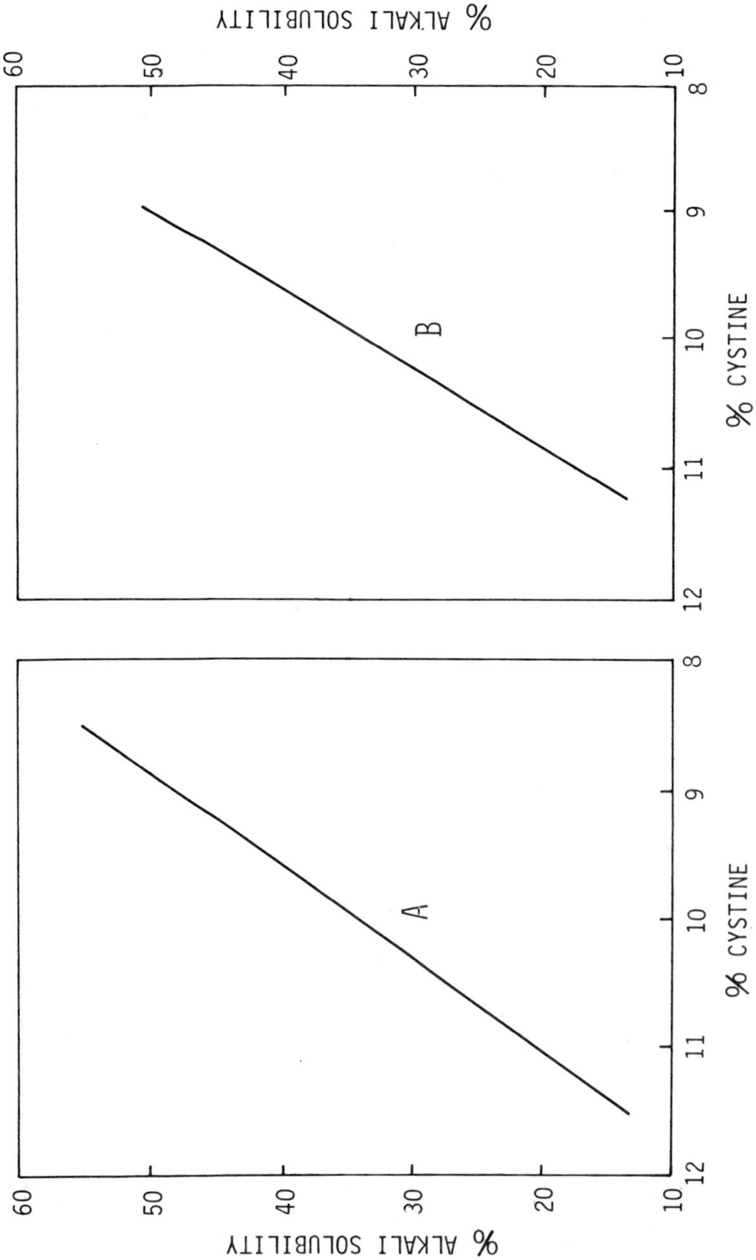

Figure 7. The alkali solubility of bleached wool in relation to the cystine content. A: Based on data from Ref. 28. B: Redrawn from Ref. 29.

hardly affected by hydrogen peroxide treatment although a significant increase in alkali solubility indicated that extensive chemical changes occurred during the treatment. G. Blankenburg (35) confirmed this unexpected behaviour of wool bleached under a wide range of conditions. How can we explain this discrepancy? The Harris and Brown data were obtained with reduced and S-methylated wool samples, in which the disulfide bonds were permanently disrupted. In oxidized wool, on the other hand, two fully ionized sulfonic acid groups are produced from each disulfide link. These groups react with the salt links in wool forming new strong salt linkages of the ammonium-sulfate-type (36)(Figure 8).

If one postulates a higher strength of such sulfonic acid-ammonium bonds in oxidized wool, one refers again to classical work of the Harris school. J. Steinhardt, H. Fuggit and M. Harris (37) showed that the quantity of hydrogen ion combined at a given pH is highly dependent on the nature of the anion. The more complex the anion the greater the affinity of the acid for the fiber. Combined cysteic acid can be considered to deliver a more complex anion than combined glutamic acid. This hypothesis might explain the intact wet strength of bleached wool fibers. There is no contradiction with the greatly increased alkali solubility of bleached wool. This can be understood on the basis of increased osmotic swelling (Figure 9) following the dissociation of the COOH groups (Wolfram et al.)(38) and rupture of the cysteic acid salt links at the onset of dissociation of the $NH_3^{\oplus}$-groups of combined lysine and of the guanidinium groups of combined arginine.

Felting and its prevention

"The tendency of wool and other animal fibers to felt constitutes one of the most unique and important properties possessed by any class of fibers. On the one hand it has contributed in a major way to the successful use of wool in a great variety of clothing and industrial fabrics. In contrast, it often reduces the usefulness of wool in items for which this fiber is admirably suited but which require frequent laundering". D. Frishman and M. Harris (39) speak concisely in support of the necessity of doing research on the felting as well as the shrink-resistance of wool, which they in fact carried out over a period of many years. It was Harris and coworkers (40) who developed a simple quantitative test for the felting of wool tops. This was the first time that the feltability of wool tops could be determined quantitatively. Earlier tests were almost exclusively done on knitted samples. The "top shrinkage test" provided a useful tool for controlling the application of antishrink treatments. Due to the fact that the Hercosett antishrink

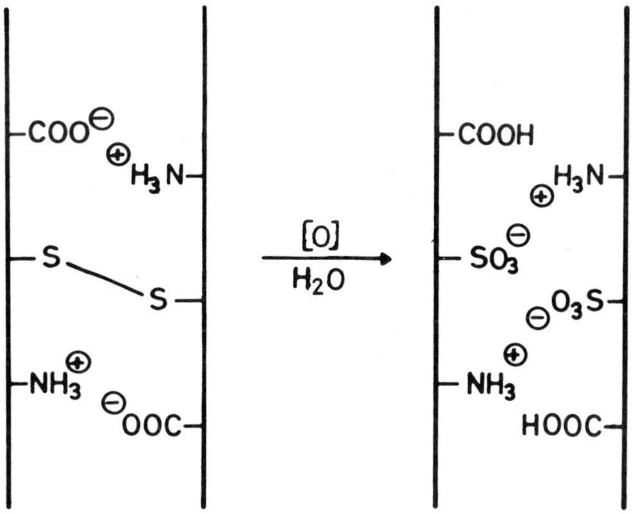

Figure 8. Rearrangement of salt links in wool during oxidation. A: Schematic representation of one disulfide and two salt cross-links in untreated wool. B: Oxidation converts the disulfide link into two cysteic acid residues which react with salt linkages. Carboxyl groups of combined glutamic and aspartic acid are set free.

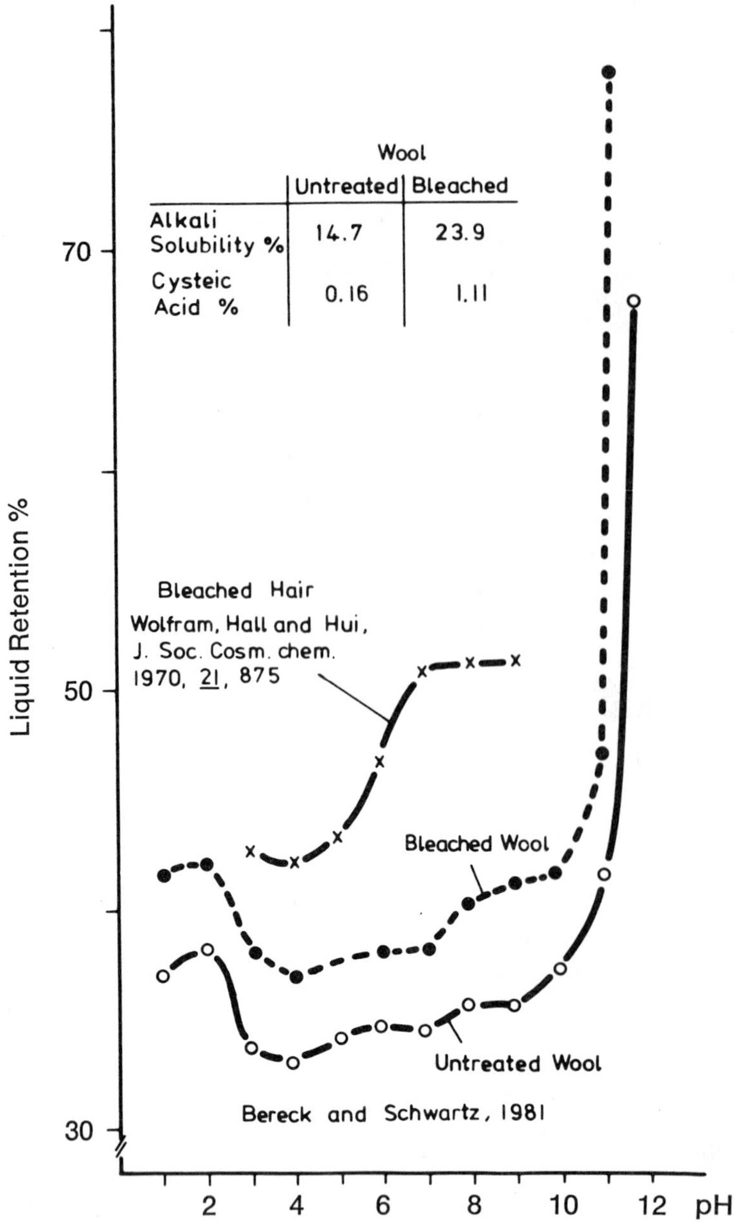

Figure 9. Liquid retention of untreated and bleached wool as a function of pH (30). Data for bleached hair are redrawn from Ref. 38.

process on tops is nowadays the one with the widest application, the "top shrinkage test" has kept its great importance for the wool industry.

Stimulated by the work of Harris and coworkers, G. Blankenburg (14) developed the "Aachen Filztest", the so-called "feltball test". In this test 1g loose wool or top is shaken in a shaking machine with a 3-dimensional motion. In this way feltballs are formed. Their diameters are directly correlated with the felting capacity of the wool. This test has for some time been accepted as a standard by the International Wool Textile Organization. Figure 10 shows the correlation between top shrinkage and felt ball density and the pH of the felting medium. The general shape of the curves are very similar.

H. Bogaty, L.I. Weiner, A.M. Sookne and M. Harris (41) were the first who acknowledged the importance of knitting density for laundering shrinkage. Different types of knitted fabrics were washed in a laundering machine. They found that an increase in the knitting density (cover factor) resulted in a decrease of the area shrinkage of the fabrics. A linear relationship was found between these two parameters. 12 years later H.J. Henning (42) carried out very similar work. Using a rotating drum type washing test he also found this linear relationship between "cover factor" of the knitted fabrics and the area shrinkage in washing. Figure 11 was taken from Harris' work (41) and Figure 11b from Henning's publication (42). The close agreement between the two diagrams can be seen.

Let us now turn to the development of shrink-resistant treatments. Up to 100 processes have been described for the production of shrink-resistant wool, and of these about half involve the use of chlorine. D. Frishman and M. Harris (39) postulated that the reaction of chlorine should be limited to the surface of the fiber since chlorine produces shrink resistance by changing the surface frictional properties of the wool fibers. The Harriset-process was one of the most sucessful continuous proccesses and the work of Harris certainly has been fruitful, since the continuous treatment developed by C.S.I.R.O. for applying synthetic polymers to pretreated wool tops (43)is to-day the leading continuous process for wool top antishrink treatment. The latest continuous process seems to be the Kroy method whereby chlorine dissolved in water acts for very short time on wool tops as a pretreatment for the subsequent polymer deposition.

As far as the chemistry involved is concerned, it was again Milton Harris and his school (44, 46) who contributed essentially to the understanding of the reaction of chlorine and its derivatives with wool. It must suffice to enumerate some of these contributions namely the formation of intermediate oxidation products of cystine residues, the maximum modification of cystine at pH 4-5, the partial dissolution of wool proteins in the chlorine solution and its pH dependence (Figure 12), the

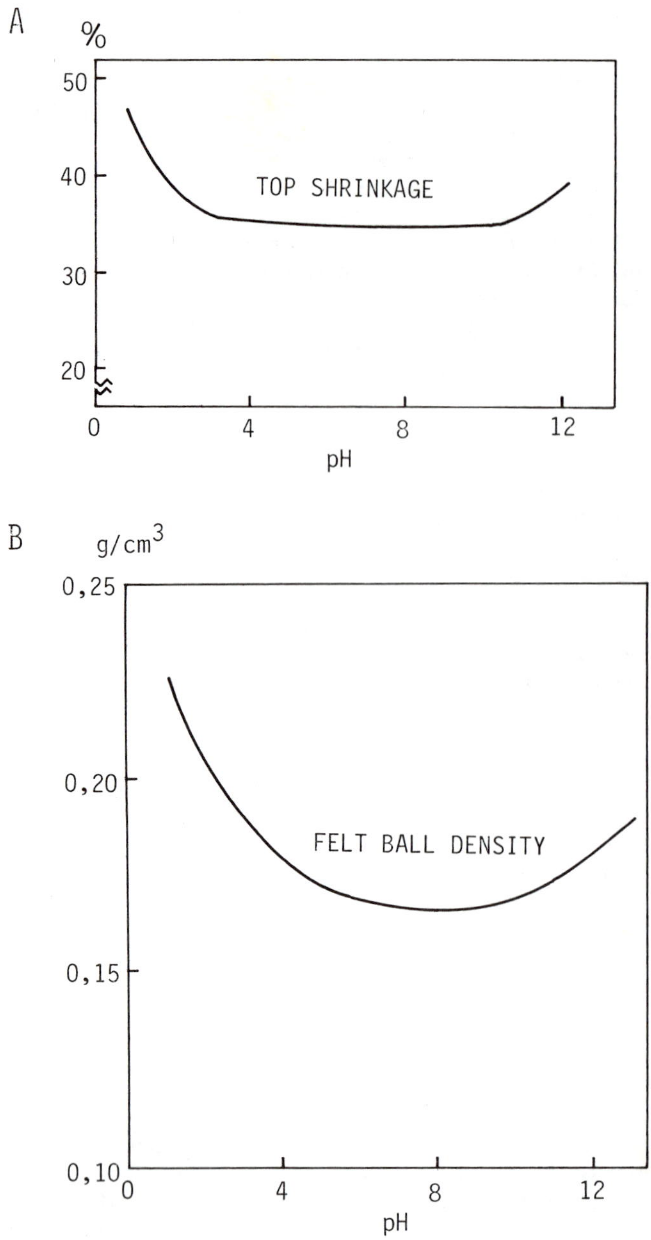

Figure 10. A: The effect of medium pH on top shrinkage (40). B: The effect of medium pH on felt ball density (14).

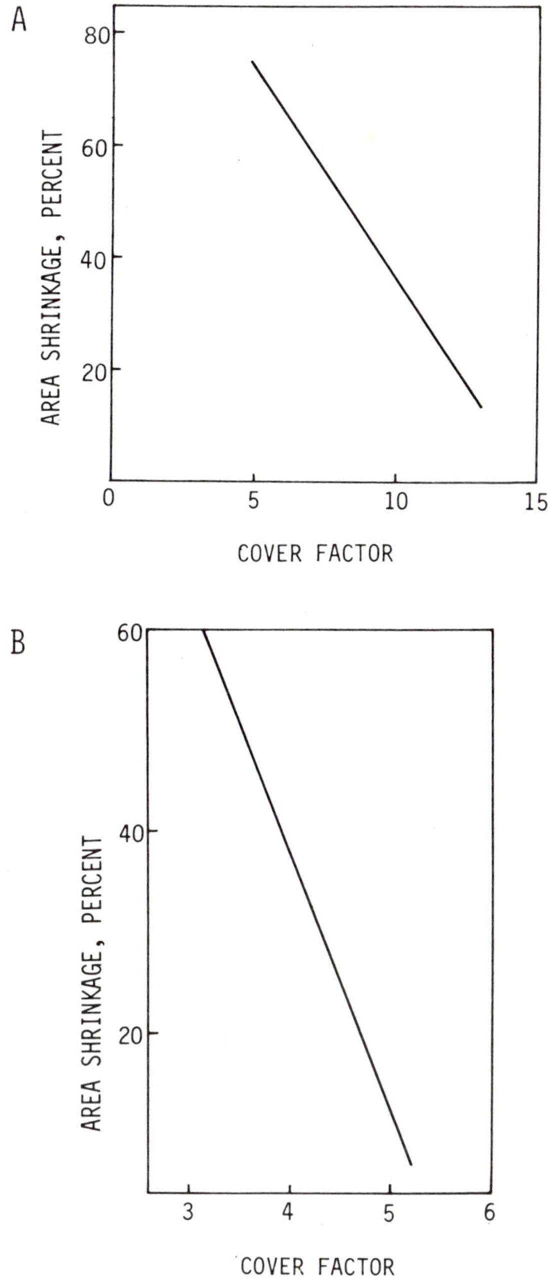

Figure 11. A: Shrinkage of untreated flat-knit fabrics in laundering as a function of cover factor (41). B: Area shrinkage of knit fabrics as a function of cover factor (42).

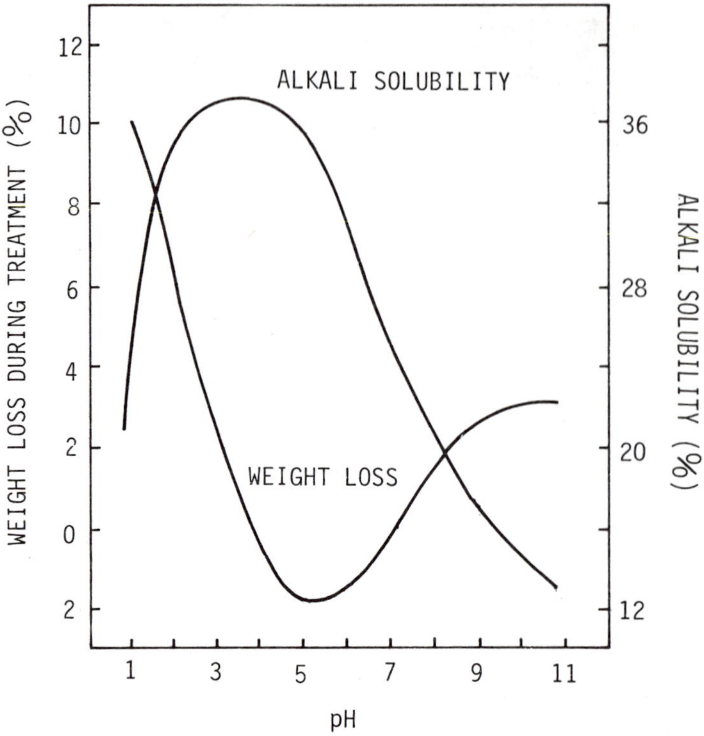

Figure 12. Weight loss during chlorination and alkali solubility as a function of hypochlorite treatment solution pH (46).

slower reaction rate of hypochlorite compared to that of either hypochlorous acid or free chlorine. In addition Harris and Smith introduced their alkali solubility test which has already been mentioned as a means for assessing the degree of chlorination.

We make full use of all the data given by Milton Harris. During rereading of many of his publications, we came across his test for the quantitative assessment of residual oxidizing agents or intermediate oxidation products of cystine, based on treating the wool sample with HI and titrating the liberated iodine. Prömper et al. (45) have rediscovered the wide application range of this test for detecting and measuring residual chlorine, peroxide and other oxidizing agents in chlorinated or bleached wool samples.

One other point of great commercial interest is the weight loss involved in wet chlorination. Again thanks are due to Milton Harris for having investigated this facet of the chlorination oxidation of wool.

Blankenburg and Töpert (47) only recently found that in the continuous chlorination process on a back-washing machine weight losses of the order of 3% occur. If from these chlorinated fibers ultrathin sections are made, stained with silver nitrate and studied in the electron microscope, as has been done by Blankenburg and Kassenbeck (48), it can be seen that only the outer, cystine-rich layers of the cuticle are oxidized and removed from the fiber. No alteration of the cortex or of the cystine-poor endocuticle could be observed. Calculations, made from the electron microscope photos, came also to weight losses of the wool close to 3%, as was found earlier by weighing.

Dyeing in the Presence of Anionic Surfactants

The contributions of Harris and his school to the theory and practice of wool dyeing comprise one important part of his work. There is even today no thorough review on wool possible that does not take into account the results of the titration of wool with acids and alkalies, the affinities of anions of inorganic and organic acids, especially the acid dyestuffs, the astonishingly high uptake of ionic surfactants, the effect of salts in increasing the uptake of Orange I. We were greatly influenced in our work by the paper of J. Steinhardt and C.H. Fugitt (49) on the relative effectiveness of various strong acids in hydrolyzing the primary amide bonds of glutamine and asparagine in wool. They found that the sodium salts of long chain alkyl sulfonates catalyse the hydrolysis of wool by hydrochloric acid. They estimated the hydrolysis of amide bonds by measuring the ammonia evolved (Figure 13). In the meantime, protective agents for wool in acid dyebaths have been used based on alkyl sulfate and sulfonate surfactants.

In an attempt to clarify these contradictions, Meichel-

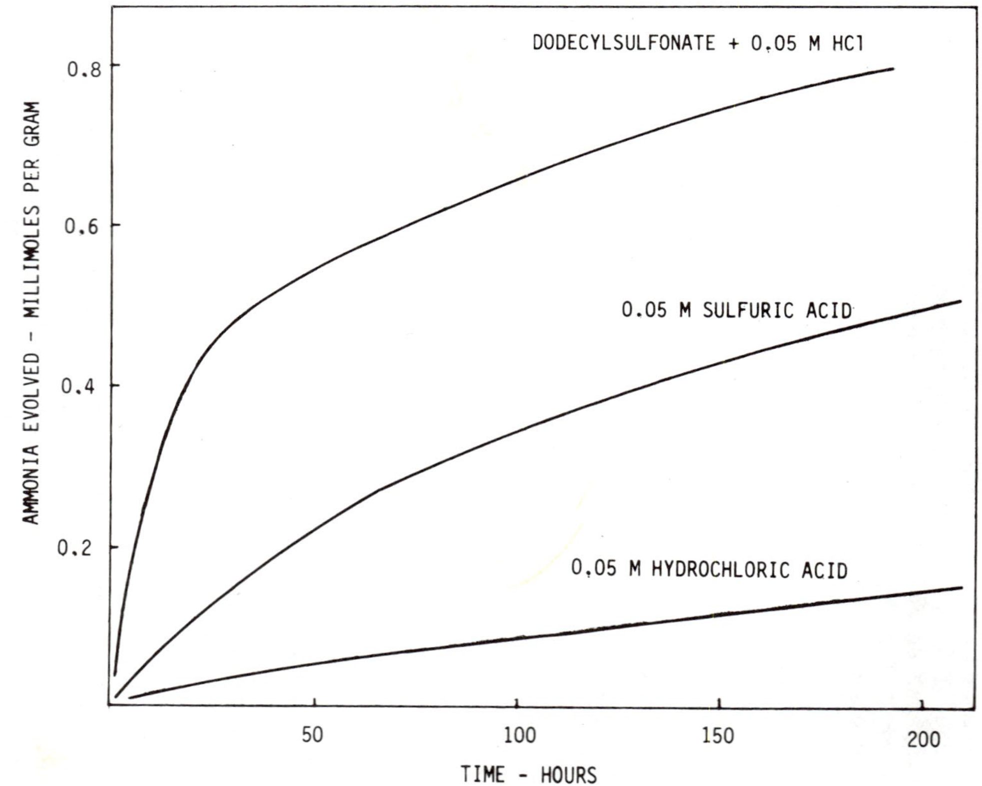

Figure 13. Relative effectiveness of strong acids in hydrolyzing the primary amide bonds of wool (49). The acids are compared at the same concentration of hydrogen ion (0.05 M) and temperature (65°C).

beck and Knittel (50) investigated the effect of sodium dodecyl sulfate on wool damage under conditions nearer to those used in practical dyeings (pH 3, 100°C, 1 h). They used alkaline solubility as the criterion for wool modification. Damage to the wool was found to increase greatly in the presence of the surfactant and more rapidly at surfactant concentrations above 10^{-3} molar.

E. Finnimore, J. Knott and G. Töpert (51) reinvestigated the results of Meichelbeck and Knittel but using changes in the (α+ε) amino group content of the wool rather than alkali solubility as the criterion for wool modification. This allows a direct estimation of the amount of peptide bond hydrolysis which has occurred during the treatment.

The effect of a pure anionic surfactant and two commercial products on wool modification under acid dyeing conditions has been investigated. The uptake of anionic surfactant was shown to have a catalytic effect on hydrolysis as shown by Figure 14.

During rereading of the paper by J. Steinhardt and C.H. Fugitt (49) it became obvious that the estimation of the amide-nitrogen content of wool could serve as a means for estimating the chemical integrity of the wool proteins. Under hydrolytic conditions during dyeing and wet finishing it should always be the amide bonds in the side chains which are attacked faster than the peptide bonds within the main chain. The estimation of the amide-nitrogen is well known in the hide processing industry and offers the advantage of being simple and unaffected by most textile chemicals including dyestuffs. Ingrid Steenken (52) has therefore applied this test for estimating the attack of alkalies on glutamine and asparagine side chains in wool. The amide content fell even under the mild conditions of a short treatment of wool with sodium carbonate solutions at 40° for 60 minutes. There was a good correlation with increased liquid retention as is shown in Figure 15.

Crease Recovery-Wrinkling

The specification of the performance characteristics of wool fabrics has proved to be a very difficult problem. One particular, yet very important and complex, performance problem in wool fabrics is that of wrinkling.

Wrinkling, especially in light-weight suiting materials, causes many complaints by consumers. The problem has been studied for many years all over the world and the publication by J.F. Krasny and A.M. Sookne (53) describes the impact of fiber properties on the crease and wrinkle recovery. "A fair correlation was found between fabric crease recovery on the one hand, and fiber tensile and work recovery on the other hand at low and normal humidities. However, the possibility of other

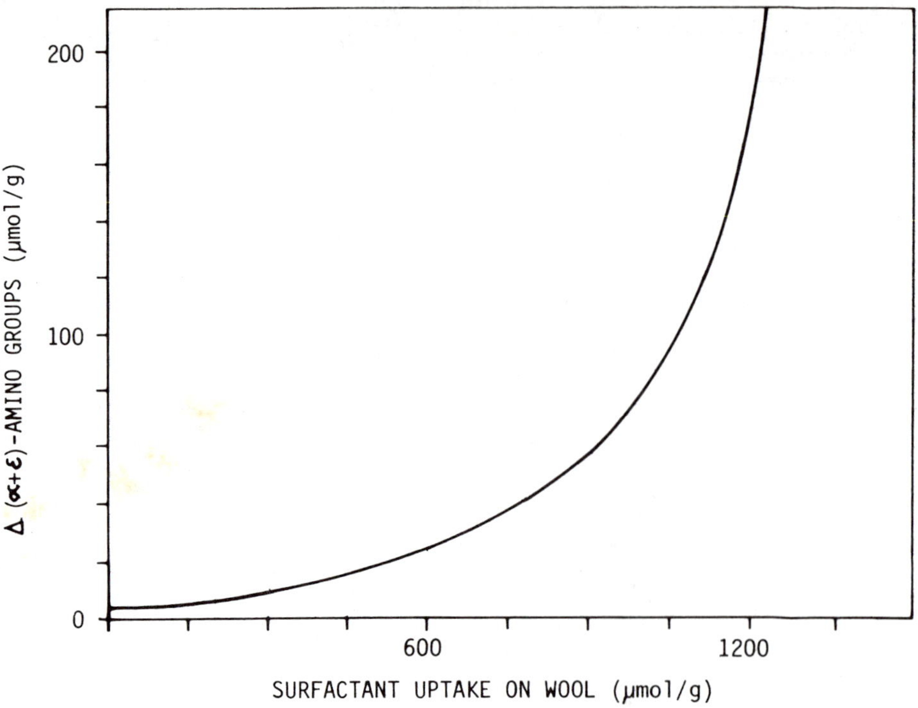

Figure 14. Blank dyeing of wool with sodium dodecyl sulfonate (pH 2, 100°C, 1 h, liquor ratio 1:360). Primary amino group content change (α + ε) in micromoles per gram in relation to the uptake of surfactant in micromoles per gram (51).

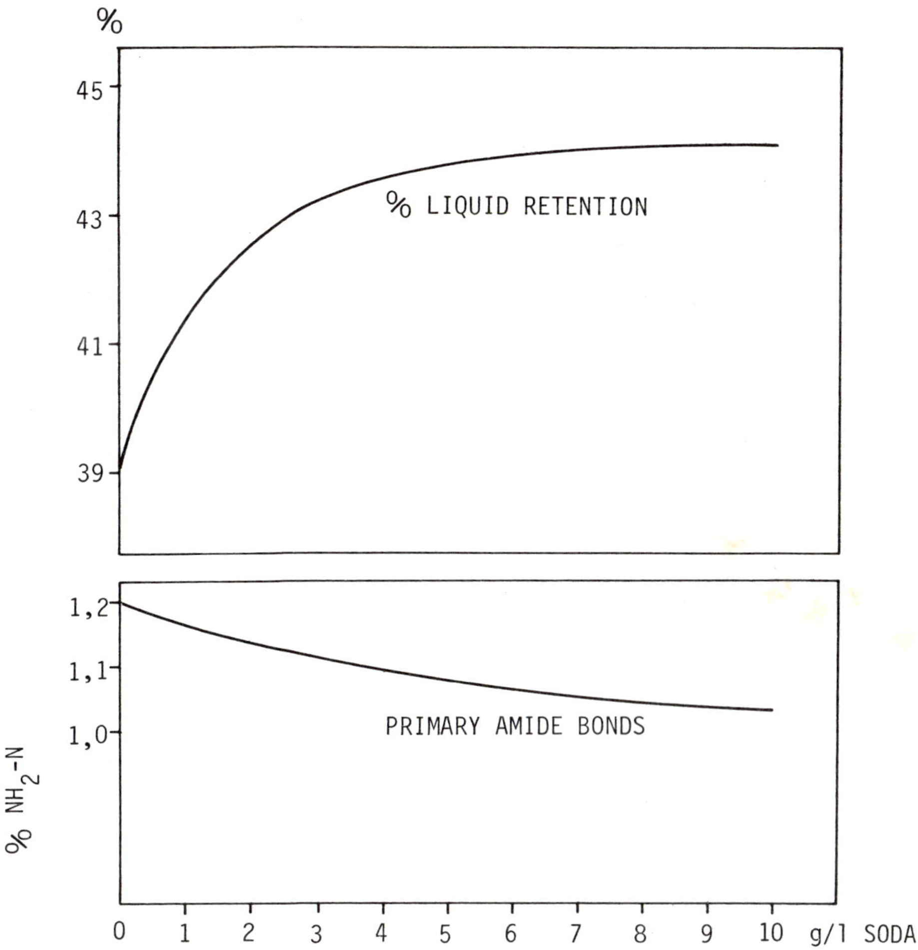

Figure 15. Reaction of wool with sodium carbonate solutions at 40° for 60 min. Liquid retention and concentration of primary amide nitrogen in wool as a function of sodium carbonate concentration (52).

fiber properties correlating better with crease recovery should not be discounted. The question of which parameters have the greatest effect on fabric crease recovery remains unanswered." This quotation can be regarded as the link to the search to find these parameters as part of an intercontinental wool research program of studies of wool fabric performance and its relationship to the viscoelastic nature and sorption properties of the wool fiber. The project involves contributions from the University of New South Wales, C.S.I.R.O. and the German Wool Research Institute in Aachen.

Clothing physiology

Physiological data is being used more and more as an aid to selling new fibers and new products and to attack the position of established natural fibers such as wool and cotton. Do the wool research and market people know how much work has already been published by Harris and coworkers in the field of clothing physiology? Do they know the publication by C.W. Hock, A.M. Sookne and M. Harris (54) on thermal properties of moist fabrics?

The results of their experiments showed clearly the progressive improvement of the fabrics with respect to chilling, as their wool content is increased, and also the superiority of certain types of construction which minimize the extent of contact of the fabrics with the skin. There are many other pertinent experiments on the diffusion of water vapor through textile, rate of drying of fabrics, evaluation of the nature of a fabric surface by a rate-of-cooling method, water transport mechanisms. We have found that these fundamental papers have only occasionally been quoted in the later literature.

Epilogue

It must be stressed once more at this point that the author made his choice of subjects dealt with above quite voluntarily and following his personal interests rather than considering any kind of evaluation as to scientific impact. He apologizes for important parts of the work passing by unmentioned and would thank other speakers at the symposium especially Arnold M. Sookne for drawing attention to some of those topics.

He hopes that he has achieved his purpose which was to prove how much the Harris School has to offer to those scientists who are willing to consider it, ie. to read its publications. And this reminds the author of the banquet in 1970 at Berkely on the occasion of the Fourth International Wool Textile Conference when Milton put the question before the Assembly: do chemists listen? having spread out before them

them a bird's eye view of the enormous ecological problems and subsequent laws imminent already then. His question could easily be extended by asking: do chemists read enough? And if they do, is it the best choice they are making, ie. do they take note of publications that are based on experimental work and can be repeated and checked in the lab later on? None of us knows. But wait and see: The future development of wool research will show up clearly to what extent its promoters have been able to take up Milton Harris' ideas and incorporate his thoughts into their work.

Acknowledgments

The author wishes to thank Professors H.J. Henning, G. Blankenburg and J. Knott, Drs. E. Finnimore, A. Bereck and F.J. Wortmann of German Wool Research Institute for advice in connection with the preparation of the manuscript.

Literature Cited

1. Horio, M.; Kondo, T. Textile Res. J. 1953, 23, 373-86
2. Hock, Ch. W.; Ramsay, R.C.; Harris, M. Am Dyest. Rep. 1941, 30, 449-56, 469-70.
3. Mercer, E.H. Biochem. Biophys. Acta 1954, 15, 293-5.
4. Kusch, P.; Arns, W. Schriftenreihe des Deutschen Wollforschungsinstituts 1982, 87, 386-90
5. Fourt, L. Textile Res. J. 1966, 36, 899-902.
6. Geiger, W. Textile Res. J. 1944, 14, 82-5.
7. Nienhaus, M.; Föhles, J. Proceedings 6th Quinquennial International Wool Textile Research Conference Pretoria 1980, II, 487-95.
8. Bradbury, J.H.; Leeder, J.D.; Watt, I.C. Appl. Polym. Symp. 1971, 18, 227-36
9. Swift, A.; Bews, B. J. Soc. Cosmet. Chem. 1974, 25, 355-66.
10. Patterson, W.I.; Geiger, W.B.; Mizell, L.R.; Harris, M. Text. Res. J. 1941, 11, 379-93.
11. White, F.H. J. Biol. Chem. 1961, 236, 1353-60.
12. Loebner, C.H. "Flecken in Wollenwaren"; Grünberg in Schlesien 1889; p 132.
13. Harris, M.; Smith, A.L. J. Research Nat. Bur. Standards 1938, 20, 563-9.
14. Blankenburg, G. "Chemische und physikalische Eigenschaften von unveränderter und veränderter Wolle in Beziehung zum Filzvermögen"; Forschungsberichte Land Nordrhein-Westfalen Nr. 1154, Westdeutscher Verlag: Köln und Opladen, 1963; p 63.

15. Klee, D.; Altenhofen, U.; Zahn, H. Schriftenreihe des Deutschen Wollforschungsinstituts. 1982, 87, 68-78.
16. Elliott, R.L.; Asquith, R.S.; Hobson, M.A. J. Textile Inst. 1960, 51, T 692-702.
17. Kirst, W. Melliand Textilber. 1947, 28, 394-6.
18. Elöd, E.; Zahn, H. Melliand Textilber. 1948, 29, 17-9.
19. Zahn, H.; Wollemann, B. Makromol. Chem. 1953, 10, 122-35.
20. McLaren, J.A.; Milligan, B. "Wool Science - The Chemical Reactivity of the Wool Fibre"; Science Press: Marrickville NSW, Australia, 1981; p 32.
21. Brown, A.E.; Harris, M. Ind. Eng. Chem. 1948, 40, 316-22.
22. Harris, M.; Brown, A.E. Am. Dyest. Rep. 1947, 36, 316-9.
23. Kirst, W. Melliand Textilber. 1947, 28, 314-6.
24. Brandenburg, D. "Crosslinked Insulins: Preparation, Proportion, and Applications" in Protein Crosslinking, Plenum Publishing Company, 1977; p 263.
25. Mizell, L.; Davis, A.E.; Oliva, F.C. Textile Res. J. 1962, 32, 497-505.
26. Cole, M.; Fletcher, J.C.; Gardner, K.L.; Corfield, M.C. Appl. Polymer. Symp. 1971, 18, 147-61.
27. Schmitz, I.; Baumann, H.; Zahn, H. Proc. Int. Wool Textile Res. Conf. Aachen 1975. 1976, Vol. II. 313-26.
28. Smith, A.; Harris, M. Am. Dyestuff Rep. 1936, 25, 180-2.
29. Ziegler, K. Textil-Praxis, 1962, 17, 376-80.
30. Bereck, A.; Schwartz, S. Unpublished Results.
31. Smith, A.L.; Harris, M. J. Res. Nat. Bur. Stand. 1936, 16, 309-13.
32. Harris, M.; Smith, A.L. Am Dyestuff Rep. 1937, 26, p 413-5.
33. Savige, W.E.; McLaren, J.A. in "The Chemistry of Organic Sulfur Compounds"; eds. Kharasch; Meyer, C.Y. Pergamon Press: Oxford, 1966; Vol. 2, p 367.
34. Harris, M.; Brown, A.E. Text. Res. J. 1947, 17, 323-30.
35. Blankenburg, G. "Die an Wollfaser-Bündeln durchgeführte Zugprüfung und die Verwendbarkeit der Ergebnisse für die Qualitätskontrolle"; Forschungsberichte des Landes Nordrhein-Westfalen Nr. 2249, Westdeutscher Verlag: Opladen, 1972; p 56.
36. Thompson, E.O.P.; O'Donnell, I.J. Austral. J. Biol. Science 1959, 12, 490-9.
37. Steinhardt, J.; Fuggit, C.H.; Harris, M. Am. Dyest. Rep. 1942, 31, 77-8, 97-100.
38. Wolfram, L.J.; Hall, K.; Hui, I. J. Soc. Cosmet. Chem. 1970, 21, 875-900.
39. Frishman, D.; Harris, M. Am Dyest. Rep. 1954, 43, 174-7.
40. Bogaty, H.; Frishman, D.; Sookne, A.M.; Harris, M. Text. Res. J. 1950, 20, 270-6.
41. Bogaty, H.; Weiner, L.I.; Sookne, A.M.; Harris, M. Textile Res. J. 1951, 21, 102-9.

42. Henning, H.J. Melliand Textilber. 1963, 44, 189-93.
43. Feldtman, H.D.; McPhee, J.R.; Morgan, W.V. Textile Manufacturer 1967, 93, No 1107, p 122.
44. Frishman, D.; Hornstein, L.; Smith, A.L.; Harris, M. Ind. Eng. Chem. 1948, 40, 2280-4.
45. Prömper, E.; Klee,D.; Altenhofen, U.; Zahn, H. International Wool Textile Organisation: Paris, 1982, Report 11.
46. Harris, M.; Frishman, D. American Dyest. Rep. 1948, 37, 52-6.
47. Blankenburg, G.; Töpert, G. International Wool Organisation: Paris, 1978, Report No 8.
48. Blankenburg, G.; Kassenbeck, P. Melliand Textilber. 1981, 62, 478-9.
49. Steinhardt, J.; Fugitt, C.H. J. Res. Nat. Bur. Standard 1942, 29, 315-27.
50. Meichelbeck, H.; Knittel, H. Fette, Seifen, Anstrichmittel 1971, 73, 25-9.
51. Finnimore, E.; Knott, J.; Töpert G. Melliand Textilber. 1981, 62, 480-1.
52. Steenken, I.; Altenhofen, U.; Souren, I,; Zahn, H. Schriftenreihe des Deutschen Wollforschungsinstituts 1982, 87, 55-67.
53. Krasny, J.F.; Sookne, A.M. Textile Res. J. 1955, 25, 493-8.
54. Hock, C.W.; Sookne, A.M.; Harris, M. Textile Res. J. 1944, 14, 144-9.
55. Kusch, R; Arns, W. Deutsches Wollforschungsinstitut Aachen, Briefe an die Mitglieder 1980, 92, 1.

RECEIVED March 18, 1982.

4

Carbohydrate Polymers: From Cellulose to DNA

R. H. Marchessault

Xerox Research Centre of Canada, Mississauga, Ontario L5L 1J9, Canada

"Innovation is the translation of scientific findings into useful goods and services."

Milton Harris (*1*)

The chemistry and technology of cellulose can be divided into three phases:

- I From Papyrus to Fine Cottons
- II From Payen to Ziegler-Natta
- III Biomass and Biotechnology

The first phase extends from the early history of civilization when cellulose in the form of twig ropes, reed baskets or crude textiles began to be used. The Chinese converted mulberry bark to handmade paper and the Egyptians did the same in a cruder fashion with papyrus. The industrial revolution brought this era of technology development to its highpoint when the textile mills of Europe were mass producing fine cotton fabrics for the rest of the world.

Payen's discovery to the demise of rayon and cellophane as the leading man-made fiber and film is the second phase. During this period, basic science and technology proceeded hand-in-hand. The development of pulping, was closely related to Payen's classical publication (2) wherein he established that woody tissue had a unique carbohydrate-rich fraction. Payen himself was influenced by Liebig, Pasteur, Van't Hoff and Lebel. Later the giant of early organic chemistry, Emil Fischer, laid the foundation for the organic chemistry used by Freudenberg in his classical proofs of cellulose structure by hydrolysis and acetylation studies.(3) At the same time, cellulose derivatives from nitrates to xanthates were being fashioned into new consumer goods. The industrial scientists were not oblivious to the controversies about the chain structure of cellulose, but innovations proceeded nonetheless.

8412-0740-2/82-0059$07.00/0

Thanks to Staudinger, Haworth, Freudenberg, Meyer, Mark and others, the chemical and physical understanding of cellulose was established by the early twenties. Then the stage was set for improved chemical cellulose and eventually dissolving pulps. Regenerated cellulose hit its peak in the 1950's and 1960's. Cellulose plastics, film and cellulose gums were also in good demand. However, once the macromolecular concepts were established, competition was bound to come. Cheap petrochemicals and a sound understanding of vinyl and condensation polymerization was the beginning of the end for the second phase of cellulose chemistry and technology.(4)

The eclipsing of cellulose as a plastic, as a fiber and as transparent film was accelerated by the spectacular development of Ziegler-Natta polymerization for olefins.(5) The ability to convert simple gases such as ethylene and propylene into useful thermo-plastics at atmospheric pressure was serendipity at its best. The low cost of petroleum in the post-war period totally distorted our belief that this technology would be dominant for centuries. This belief was supported by Natta's championing of stereoregular polymerization: "nature lost a monopoly" as he put it. Cellulose was no longer the only man-made fiber that was stereoregular.

Environmental factors also played a significant role in the demise. During the 1960's and 1970's pulp, rayon and cellophane plants laboured to comply with increasingly severe environmental regulations. This was a period when investment in R&D related to cellulose and paper technology was aimed at maintaining the status quo while investment in synthetics was moving new products and materials into the market place. During this period, it became clear that new investment in viscose technolgy in the West was beyond reach because of the cost of meeting anti-pollution regulations.

Phase III of cellulose chemistry and technology is the aftermath of the OPEC crisis and in step with the new biology. Renewable resources have appeal in their own right but couple this appeal with a technology rebirth for cellulose, wood chemistry, carbohydrate polymers, etc., and you get:

- Polysaccharide Vaccines (6, 7)
- Superswelling Fibers
- Organic Solvents for Cellulose (8, 9)
- Thermomechanical High-Yield Wood Pulps (10)
- Organosolv Pulping for the Forest Refinery (11)
- Exploded Wood (12) for Biomass Lignins (13)
- Fuels and Chemicals from Plant Materials (15, 16)
- Recombinant DNA Techniques (17)
- Supertrees for Reforestation (19)

These developments promise a rebirth and new interest in cellulose, wood and carbohydrate polymers. This is especially true when the emerging genetic engineering component is added to existing technology. Each of these topics would occupy a chapter in this book. Accordingly, the following sections will briefly survey these various topics which will be grouped under the following headings:

- Lignocellulosics
- Chemical Cellulose
- Carbohydrate Polymers

LIGNOCELLULOSICS

Most biosynthesized cellulose in plants is compounded with lignin and hemicelluloses. The composite is generally referred to as a lignocellulosic. Anselme Payen's classical study (2) showed that plant tissue, cotton linters, root tips, pith and ovules of the flowers of trees when purified by acid/ammonia treatment followed by extraction with water, alcohol and ether yielded a resistant fibrous substance now known as cellulose.

Similar studies on various woods using nitric acid/sodium hydroxide laid the foundation for the soon-to-develop pulping industry. The encrusting substance which he removed was designated later (1857) by Schulze as "lignin". Payen thus showed that wood was a composite, the components of which were separable. The so-called fibrous resistant part was clearly shown to be carbohydrate in nature by strict elemental analyses using methods only recently introduced by Liebig.

Wood as the prime example of a lignocellulosic is physically complex (cf. Figure 1) and chemically heterogeneous with lignin being distributed throughout the cell wall as well as being the glue between the fibers. The chemical utilization of wood has always been hampered by its heterogeneity. By comparison, the lowest grade of petroleum is far more homogeneous and readily lends itself to cracking and fractionation.

The trend in the paper industry is to make increasing use of hardwoods because the conifers are becoming less accessible. The cellulose in wood is principally found in the thickened secondary wall of the tracheids, the so-called S-2 layer (cf. Figure 1). Lignin is distributed from the middle lamella where it is most concentrated to the S-layer where it is admixed with cellulose and hemicellulose. The middle lamella can be thought of simply as glue cementing the fibers together while the cell wall lignin is a reinforcing matrix. Commercial chemical cellulose made by pulping/refining aims at removal of the lignin and part or all of the hemicellulose from the middle lamella and cell wall in order to free the fiber and leave pure high molecular weight cellulose.

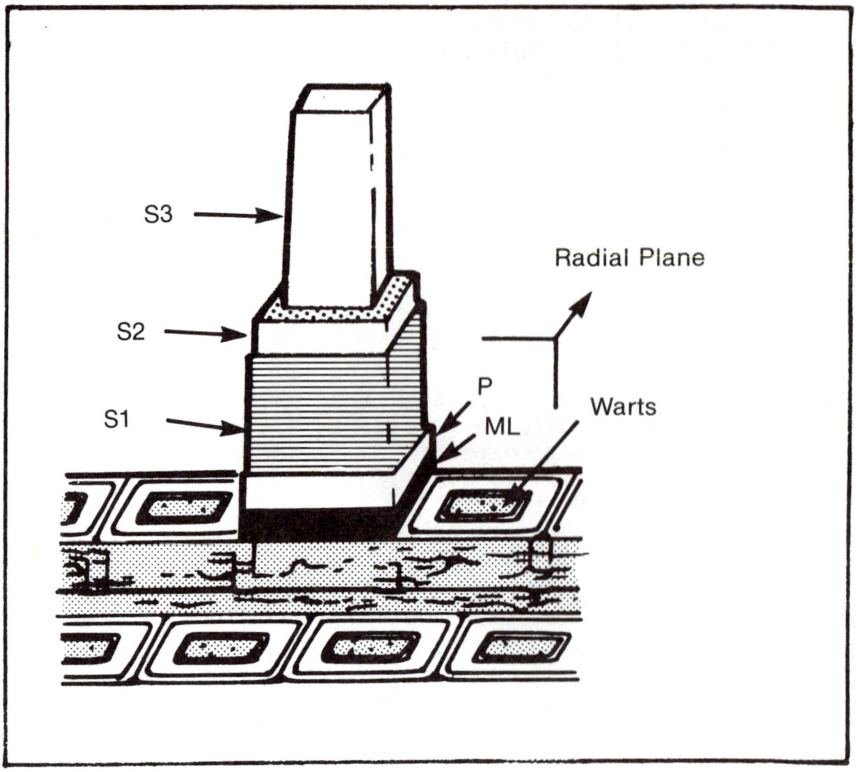

Figure 1. A schematic of cell wall structure for wood. Key: ML, *middle lamella (mainly lignin);* P, *primary wall; and* S1, S2, S3; *component sections of secondary wall differing mainly in fibrillar orientation and somewhat in hemicellulose composition.*

The presence of "encrustants" in the fibrous cells of wood is an important difference compared to cotton. Thus, pulping liquors must penetrate the fiber wall completely, but chemical attack on the cellulose must be kept to a minimum. This is aided by the relatively high crystallinity of native cellulose and the fibrillar nature of these crystalline domains.

Exploded Wood

There is need for an inexpensive and technically simple approach to unlock the nascent cellulose-lignin clasp and allow easy separation of these two major chemical components of wood. A simple approach for doing this, based on steam and temperature, has recently been demonstrated and the results of this technology provide a new approach to biomass utilization.(12) Because of the naturally acid condition of wood and the influence of the water and temperature, the process is basically one of autohydrolysis. The conditions for deciduous wood and straws may be simply stated as:

- Saturation Moisture Content of Wood
- Natural pH of the Lignocellulosic
- Rapid (50 secs - 100 secs) Temperature and Pressure Rise
- Maximum Temperature and Pressure: 200-240°C and 500-600 psi
- Rapid Decompression After Treatment

Under these conditions and given typical chip size, the product is a dark brown, moist, fibrous powder with a smell of burnt sugar and vanillin. Within the window of the above conditions, wood specialists agree that the product can be physically and chemically described as "exploded wood".

Based on the extensive and evolving wood chemistry knowledge (13, 14) the process leads to:

- Cleavage of Some Accessible Glycosidic Links
- Cleavage of β-Ether Linkages of Lignin
- Cleavage of Lignin-Carbohydrate-Complex (LCC) Bonds
- Chemical Modification of Lignin and Some Carbohydrates

The first three transformations can be demonstrated by a range of physical and chemical methods. Some of the changes such as loss of part of the methoxyl and acetyl groups are incontrovertible, others such as cleavage of LCC are speculation based on evolving concepts of wood fine structure. Loss in degree of polymerization of cellulose compared to the native state is easily established as is the breakdown of the hemicelluloses to simple carbohydrates and oligosaccharides.(13) Chemical transformations other than hydrolytic liberation of a true or pseudo-repeating unit in the linear (polysaccharide) or branched (lignin) polymers of wood can be considered unavoidable in a rapid process which takes place in a compromise which provides a window for obtaining desired physical properties in the "exploded wood". These properties are those which are important to the intense interest in biomass utilization for chemicals and energy conservation.

Extensive observations(12, 13, 14, 20) have led to the following results:

- Comminution to papermaking or hardboard size particles, i.e., partial preservation of fibrous texture.
- Preservation of cellulose crystalline structure.
- Transformation of lignin to a state where it is 96% soluble in simple solvents such as methanol, elthanol and acetone.
- Transformation of hemicelluloses to a state where they are predominantly extractable by water.
- Cellulolytic enzyme accessibility of the exploded wood equivalent to what is achieved by mechanical or other physical treatments.
- Ruminant nutritive value of hardwood and straw carbohydrate raised from 30% → 60% or higher.

Since these transformations are achievable without added chemicals, the explosion process is an attractive means to convert hardwood residues, agricultural waste and suitable plantations. The exploded wood itself should be thought of as the feedstock of a forest refinery (11) since deciduous wood does not lend itself to simple solvent extraction of the lignin. Uses for the latter include replacement of the phenol component in phenolic resins and further cracking as a source of aromatics. The schematic of Figure 2 is a flow chart of how the wood explosion process can allow fractionation of wood.

The hemicellulosic carbohydrate fraction is also extractable as a high xylose content feed by prehydrolysis prior to the explosion process. This approach has been commercialized for manufacture of xylitol in Finland.(18) Furthermore, the cellulosic fraction of exploded wood retains satisfactory fiber properties to act as a filler in a low-grade paper. By hydrolysis, chemical treatment and fermentation of the carbohydrates from biomass, a wide range of chemicals containing carbon-oxygen linkages are achievable.(11, 16)

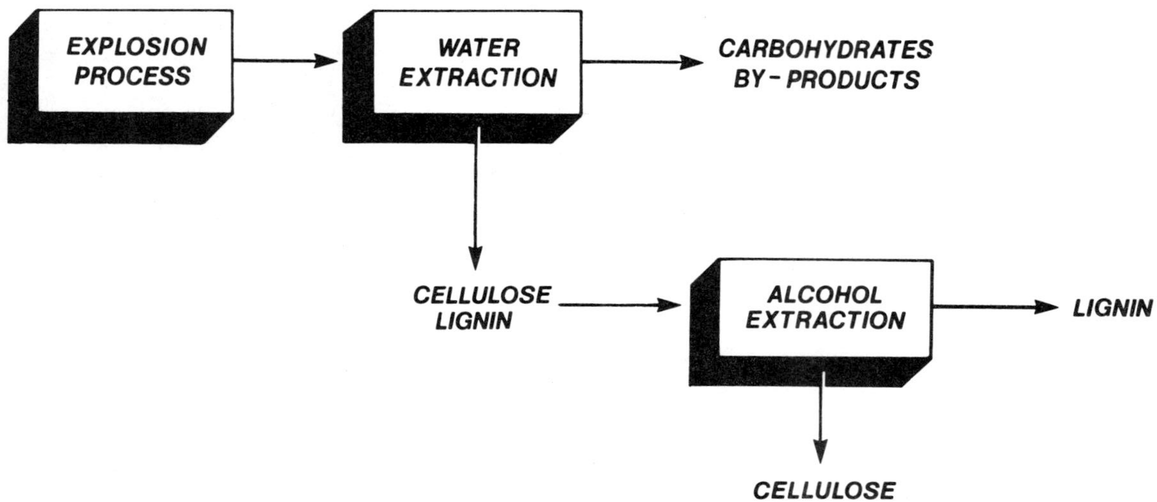

Figure 2. Flow chart of wood fractionation based on explosion process.

Chemicals from Wood

The feedstock scenarios of the chemical world require a starting material which is reasonably homogeneous. The two main chemical components of wood: carbohydrate and aromatic/hydrocarbon are so diverse that a separation must be achieved before any consideration of using wood as a feedstock. The explosion process provides a simple method for this both for hardwoods or straws whose average starting composition are respectively:

TABLE I

Average Composition* of Hardwoods and Straws in Weight %

	Hardwoods	Straws
Cellulose	45 ± 2	30 ± 5
Hemicellulose	30 ± 5	27 ± 3
Lignin	20 ± 4	11 ± 1
Extractives	5 ± 3	15 ± 5

* Reference (21)

Alcohol extraction of exploded wood provides a 20% solution of low molecular weight lignin in ethanol from which subsequent uses can flow. This is the first time that a rapid inexpensive process for separating the two components has been available.

A truly masterful analysis of the potential for chemical production from wood has recently been presented by Myerly, et.al.(11) The range of chemicals is summarized in Figure 3 which is borrowed, in part, from that study which was based on an organosolv pulping process for wood fractionation.

When one imagines the extent of process improvements which can come about as a result of recombinant DNA technologies in the future, Figure 3 is all the more impressive. In the case of exploded wood, the cellulosic substrate has greatly increased enzyme susceptibility as a result of the explosion treatment itself.(20, 22) Improvement in the activity of cellulases and hemicellulases would allow extremely rapid conversion of all the carbohydrate product of the explosion process into a fermentation substrate. Lignases coupled with pyrolysis/oxidation reactions could open a totally new range of aromatics.

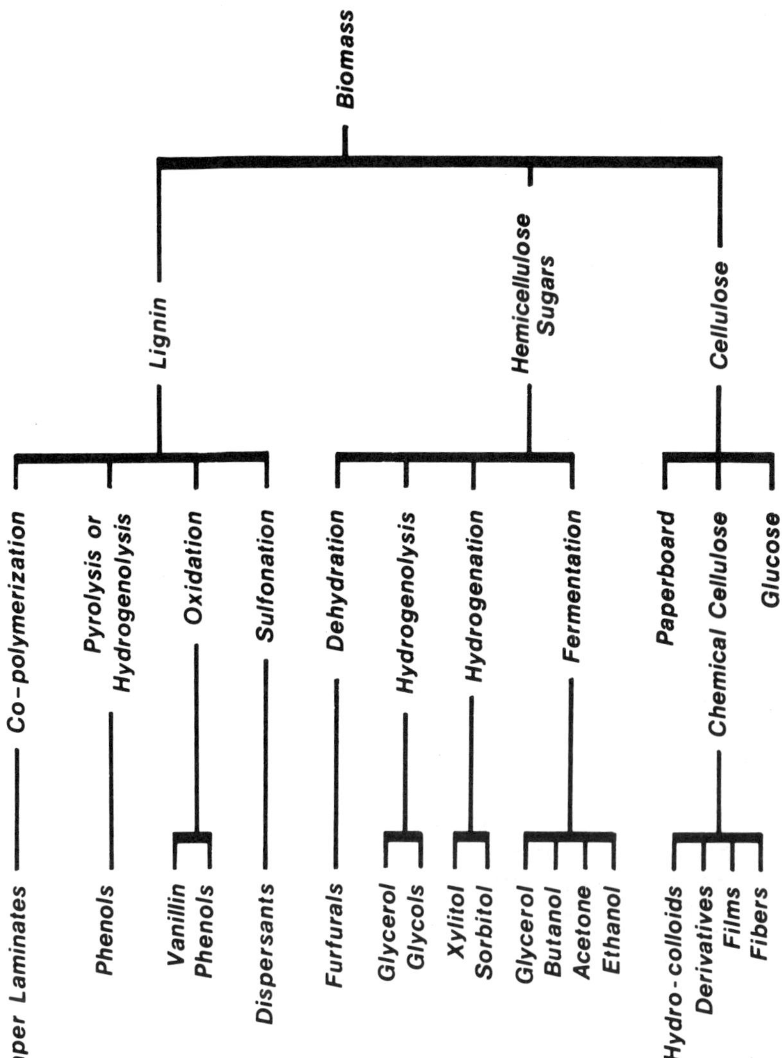

Figure 3. Chemicals from wood based on initial fractionation into lignin–carbohydrate–cellulose streams.

CHEMICAL CELLULOSE

The chemical similarities of cellulose and starch but their well-known solubility differences led some researchers in Payen's time to suggest that this was a case of the same substance in different states of aggregation. This type of natural chemistry overlapped with the development of cellulose derivatives such as cellulose nitrate which John Hyatt demonstrated in 1869, as the first useful plastic. Subsequently, man-made fibers based on cellulose xanthate (rayon) and cellulose acetate made their appearance, the key idea being to treat cellulosic matter so as to render it soluble in ordinary organic solvents.

By the late 19th century, cellulosic "man-made fibers" were an item of commerce and the first North American vicose-rayon plant of the well-known European organization, Courtaulds', was established at Marcus Hook, Pennsylvania in 1910. At that time, cotton linters were the cellulosic starting material. By the end of the 1920's, high grade wood pulp started to be used and the World War II shortage of cotton linters accelerated the change. In fact, it was the wide fluctuations in the price of linters which eventually brought about the total conversion of the rayon industry to this more practical raw material. An acre of fertile land in subtropical countries produces seven times more cellulose, as wood, than the sum of the equivalent cotton harvests over the same period.

Today, commercial cellulose from higher plant life supplies an annual world consumption of about 150 million tons of fibrous raw material. Of this amount, 60 million tons, derived mostly from pulped wood, is used for paper; about 8 million tons, mostly from cotton goes into textiles; and 3 million tons is chemical grade cellulose. The latter includes all forms of cellulose such as rayon and cellophane as well as most cellulose derivatives on the market. Chemical grade pulp contains 90-99% pure cellulose depending on process and source of raw material.

Since the technology of making dissolving grade wood pulp is so well established, there have been ongoing efforts to expand the use of chemical grade cellulose. Examples such as cellulose fibers in bread or microcrystalline cellulose for compounding pills have received extensive publicity. Two innovations will be described in the following sections: synthesis of superswelling fibers and a new solvent for fabricating rayon fibers.

Fibrous Hydrogels

Since pure cellulose fibers are not only insoluble in water, but also only slightly swollen, it is necessary to derivatize in order to approach a hydrogel character. When cross-linking and derivatization are done simultaneously, then it is possible to retain the fibrous nature and produce a hydrogel at the same time. Figure 4 shows such a transformed native cellulose fiber in a highly swollen condition. A wide range of chemical methods for achieving this transformation have been developed of which graft polymerization (23) and carboxymethylation (25) are good examples. Several commercial superswelling fibers of the latter type are available (24) which hold from 20 to 30 times their weight of water.

By retaining the fibrous character of native cellulose and imparting the swelling properties which are typical of synthetic polymer hydrogels novel textile and non-woven applications are possible. A quite unexpected development which might anticipate a unique marriage of paper and synthetic fiber technology was the conversion of superswelling fibers into one-dimensional paper.(26, 27) This was achieved by extrusion of a concentrated aqueous suspension (5-10% solids) of superswelling fibers through a millimeter-sized spinneret. The resulting superfiber or "one-dimensional paper" is a stiff whisker or soft yarn depending on drying conditions (cf. Figure 5).

It is the soft jelly-like property of individual superfibers which allows the extrusion of such concentrated suspensions. Papermakers have long wanted to go from dilute fiber suspensions to more concentrated systems for papermaking and this development clearly shows what transformation is required in the fibers in order to do so. Since the x-ray fiber diffractogram of superfibers (27), as in Figure 5, shows evidence of molecular orientation along the fiber axis, a combined laminar/extensional flow mechanism, rather than plug flow, must be operative during the extrusion of the concentrated superfiber suspensions. This is made possible by the lubricated fiber surfaces of the highly swollen partial derivatives(28).

Organic Solvent for Cellulose

Since the advent of melt spinning technology, regenerated cellulose manufacturers have sought to emulate the speed and environmental cleanliness of this fiber/film-making technology. The fact that cellulose decomposes before melting, has made this impossible. Recently, a renewed interest in cellulose solvents has resulted in the discovery of a new class of organic molecules: cyclic amine oxides which are powerful solvents for cellulose.(8, 32) Most notable among these solvents is N-methyl-morpholine-N-oxide (MMNO) which is shown in Figure 6. This molecule is reported (29) to form a monohydrate with a melting point of 72°C, the structure of which is shown in Figure 7.

The molten hydrate is a solvent for cellulose which cleaves the intermolecular hydrogen bonds of the cellulose lattice and replaces them with even stronger hydrogen bonds between the N→O group.(30, 31) The crystalline structure (30) of a

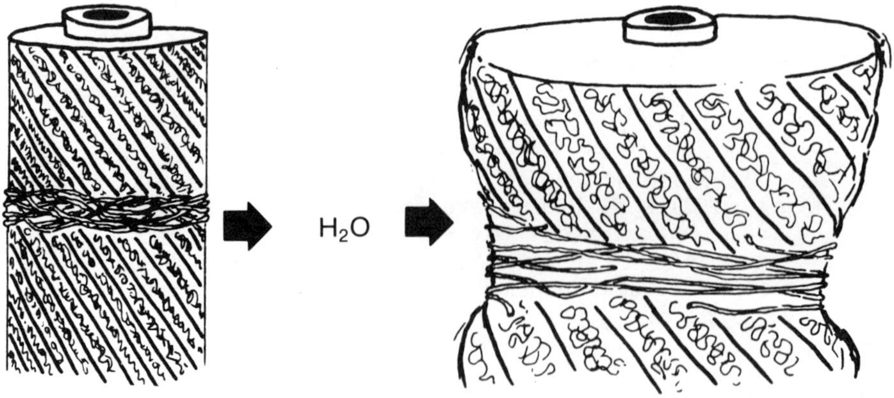

Figure 4. Schematic of copoly(acrylic acid–acrylamide) grafted cellulose fiber before and after swelling in water. Only weak mechanical and secondary forces prevent a complete dispersal into microfibrillar components.

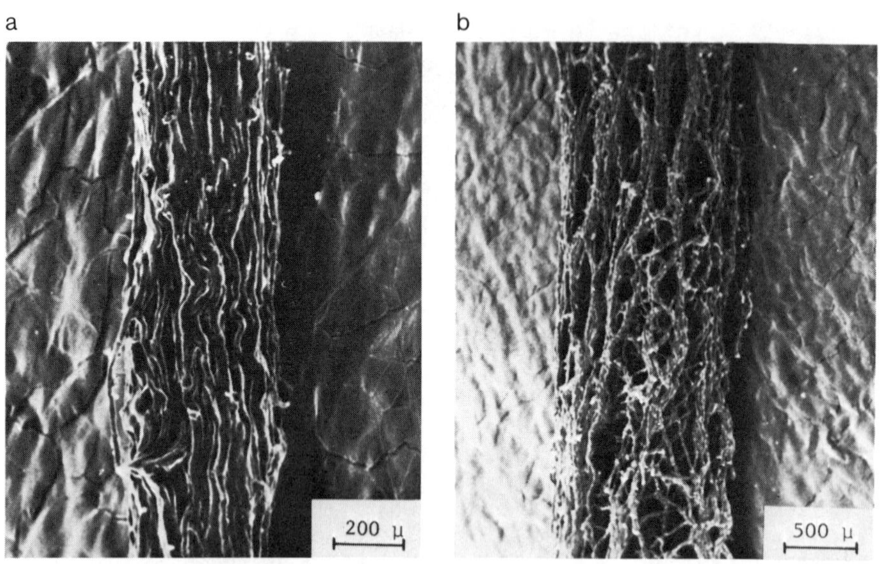

Figure 5. One-dimensional paper shown as a yarn-like product of spinneret extrusion of superswelling fibers in water. After drying, typical interfiber bonds form between cellulose fibers, as in paper.

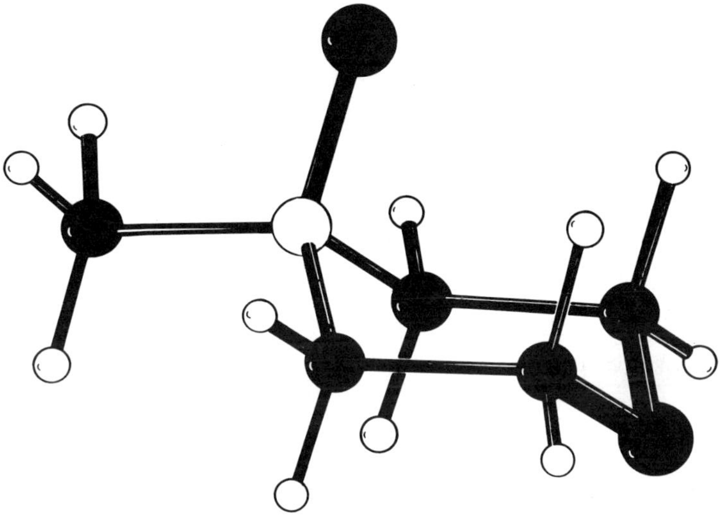

Figure 6. Conformation of N-*methylmorphine oxide.*

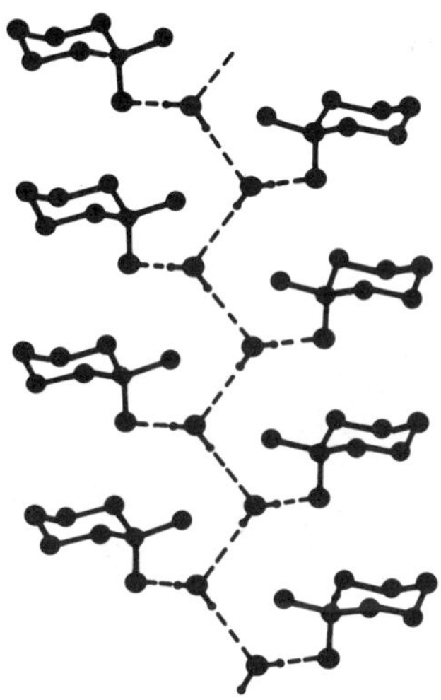

Figure 7. Crystalline organization of N-*methylmorpholine oxide with one water of hydration per molecule forming central zig-zag hydrogen-bonded column.*

model complex: cyclohexane diol and MMNO is shown in Figure 8. The diol may be replaced conceptually, with a cellulose chain which provides some notion of the dissolved state of the macromolecule.

This organic solvent for cellulose is being used to experiment with new cellulose spinning technology based on extrusion.(32) Economic success of such experiments is dependent on recovery of the solvent. The technology promises to eliminate the unfavourable environmental features of the viscose process and provide high tenacity fibers. In addition, it provides a new medium for some cellulose reactions and encourages fundamental studies of cell wall materials since the intact wall components are often soluble in the new solvent from which they can be selectively precipitated.(39)

CARBOHYDRATE POLYMERS

The wide range of properties which are found in cellulose derivatives and plant polysaccharides such as pectins, galactomannans, etc., is being duplicated and surpassed by bioconversion of starch waste to regular copolysaccharide gums. Thus, the use of fungal and microbiological systems to produce such carbohydrate polymers is on the increase. The range of properties which can be achieved is easily understood by examining the molecules in Figure 9. A range of flexibility is achieved by increasing the number dihedral angles in the chains. The potential flexibility of DNA because of the numerous dihedral angles is compensated by its ability to wind two chains into a double helix, hence greatly limiting the angular range for any dihedral angle. Similarly, regular copolysaccharides can sometimes form double or even triple helices, thereby mimicking the classical grammar of molecular biology.

It has become established that nature does take advantage of the large number of structures which can be formed from a relatively small number of carbohydrate monomers in order to maintain structure or to allow "recognition" and specific complexation. The rules for structure specificity are based on:

- The nature of the sugars present.
- The position and configuration of the glycoside linkage.
- The presence or absence of substituents.

The numerous examples of regular complex copolysaccharides often involve familiar-looking material for cellulose chemists. Figure 10 shows two pneumococcal polysaccharides, Types III and VIII, also known as "specific soluble substance", which in the 1920's and early 1930's were shown to be antigenic although they were free of nitrogen and did not possess any of the properties

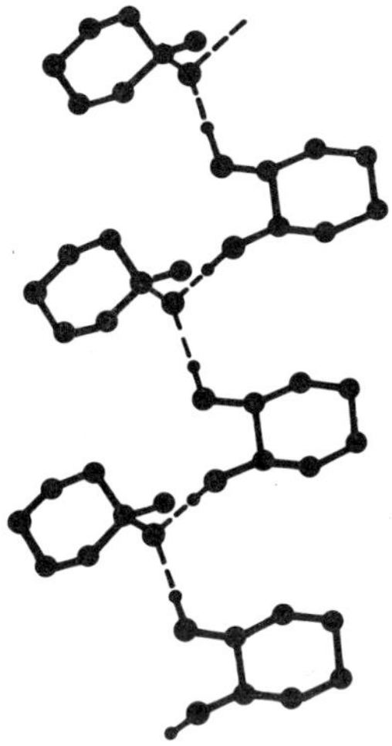

Figure 8. Crystalline organization of model complex of N-*methylmorpholine oxide and cyclohexanediol.*

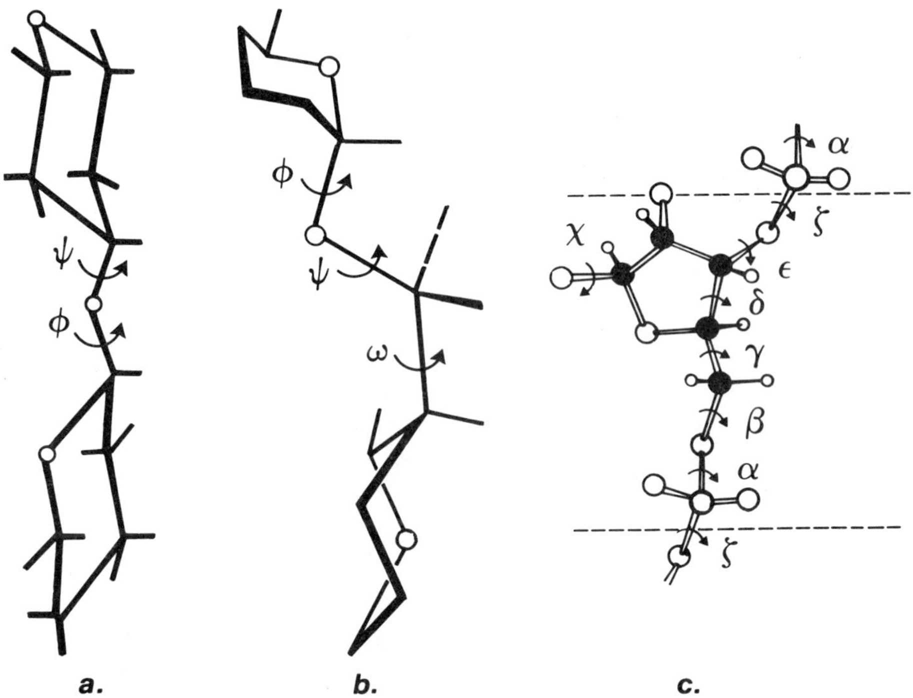

Figure 9. The flexibility and properties of carbohydrate polymers are controlled by the number of dihedral angles per repeat unit. a, β(1 → 3) glycan disaccharide; b, α(1 → 6) glycan; and c, DNA.

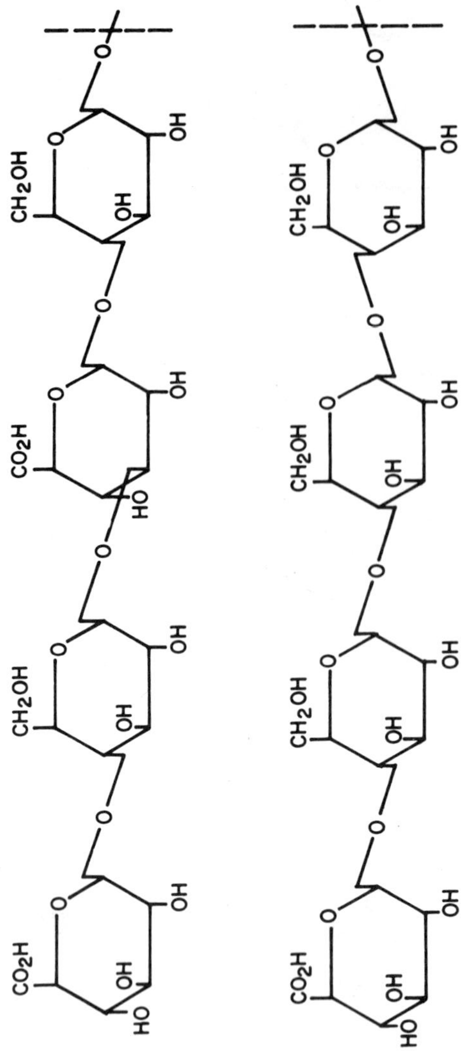

Figure 10. Chemical structure of two pneumococcal polysaccharides (regular copolysaccharides) composed of glucuronic acid and glucose residues. Differences in glycosidic configuration sequence and linkage position build in conformational specificity of immunological significance. Top, type III; bottom, type VIII.

of peptides. The knowledge achieved by the extensive studies on cellulose and carbohydrates in the first decades of this century was responsible for the early establishment of the chemical structures of Types III and VIII. The revolutionary work on bacterial transformation, in which Avery, MacLeod and McCarty in 1944 identified DNA as the genetic material (6), was done with pneumococci, the genetic marker being the production of a "specific capsular polysaccharide".(7) Thus, these materials can claim to have been on the ground floor of the development of molecular biology and have since come into use as commercial polysaccharide vaccines.

Microbiology has allowed polysaccharide gum producers to develop very unusual molecules which duplicate and surpass some of the plant polysaccharide "stars". For example, xanthan gum in Formula I has a cellulose backbone and regularly occurring carboxylated side chains. Unlike cellulose, individual chains are soluble in water, but do not show the typical polyelectrolyte viscosity effects.(33)

This type of molecule is a regular copolysaccharide not a statistical one like partial derivatives of cellulose and plant gums. While the regularly occurring side chain renders xanthan gum so different from cellulose, the influence of a different glycosidic linkage is the most important fact in establishing chain conformation.(34, 35) Figure 11 shows the minimum energy conformations of cellulose and β(1→3) glucan. The ribbon-like cellulose conformation is transformed to a large amplitude helix by a simple change of the positional linkage.

This conformational difference has far-reaching effects on the crystallization behaviour. Thus β(1→3) glucans form triple helical structures and crystallize in the triplex form (36) as shown in Figure 12. This triplex persists to varying degrees in solution, as with collagen/gelatin and leads to remarkable viscosity (37) effects within families of β(1→3) glucans. Commercially, these various species are known as curdlan, the pure β(1→3) glucan and scleroglucan (33) which is a regular copolymer with a tetrasaccharide repeating unit as shown in Formula II.

$$\begin{array}{c} \beta\text{-D-Glc}p \\ 1 \\ \uparrow \\ 6 \\ -\left[\rightarrow 3)\text{-}\beta\text{-D-Glc}p(1\rightarrow 3)\text{-}\beta\text{-D-Glc}p(1\rightarrow 3)\text{-}\beta\text{-D-Glc}p(1\text{-}\right]_n\rightarrow \end{array} \qquad \text{(II)}$$

While the former is only alkali soluble and maintains its triplex structure in water, the latter dissolves readily in water and maintains an ordered conformation in this solvent. As a result, its pseudoplastic solutions have good viscosity tolerance to a broad range of temperature, pH and salt concentrations.

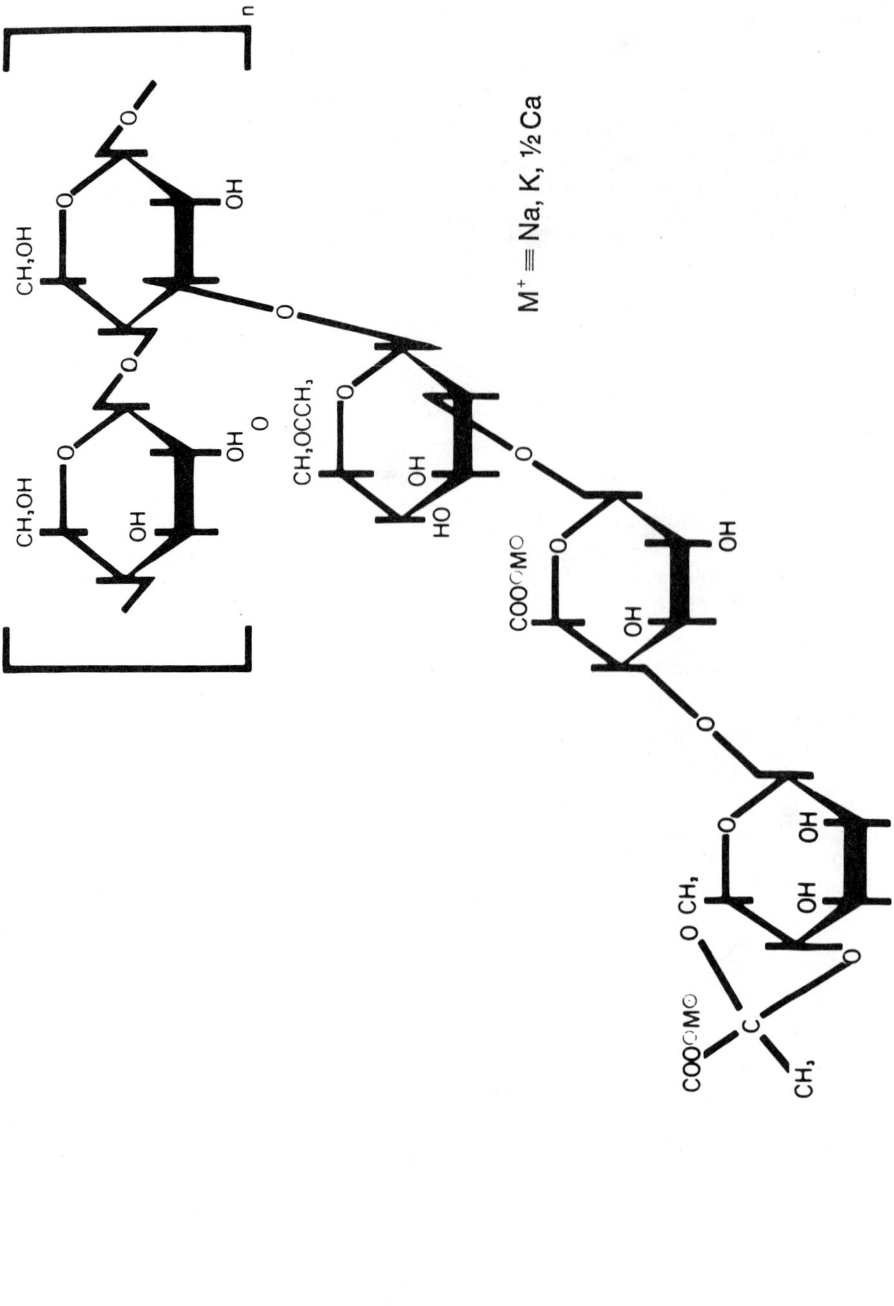

Formula I.

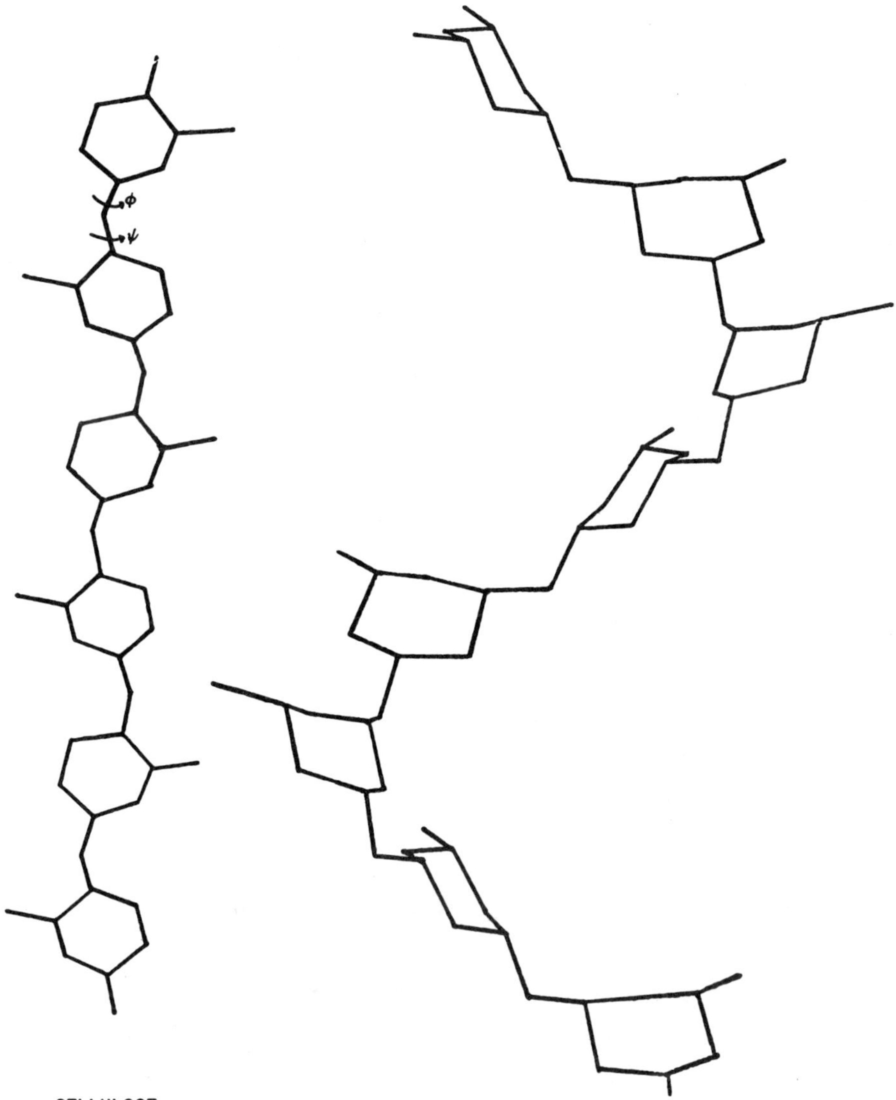

Figure 11. Cellulose, (1 → 4)-β-glucan, and (1 → 3)-β-glucan in their minimum energy conformations. The ribbon-like shape of cellulose is contrasted with the large open helix of (1→ 3)-β-glucan. The glysocidic dihedral angles φ and ψ are indicated for cellulose.

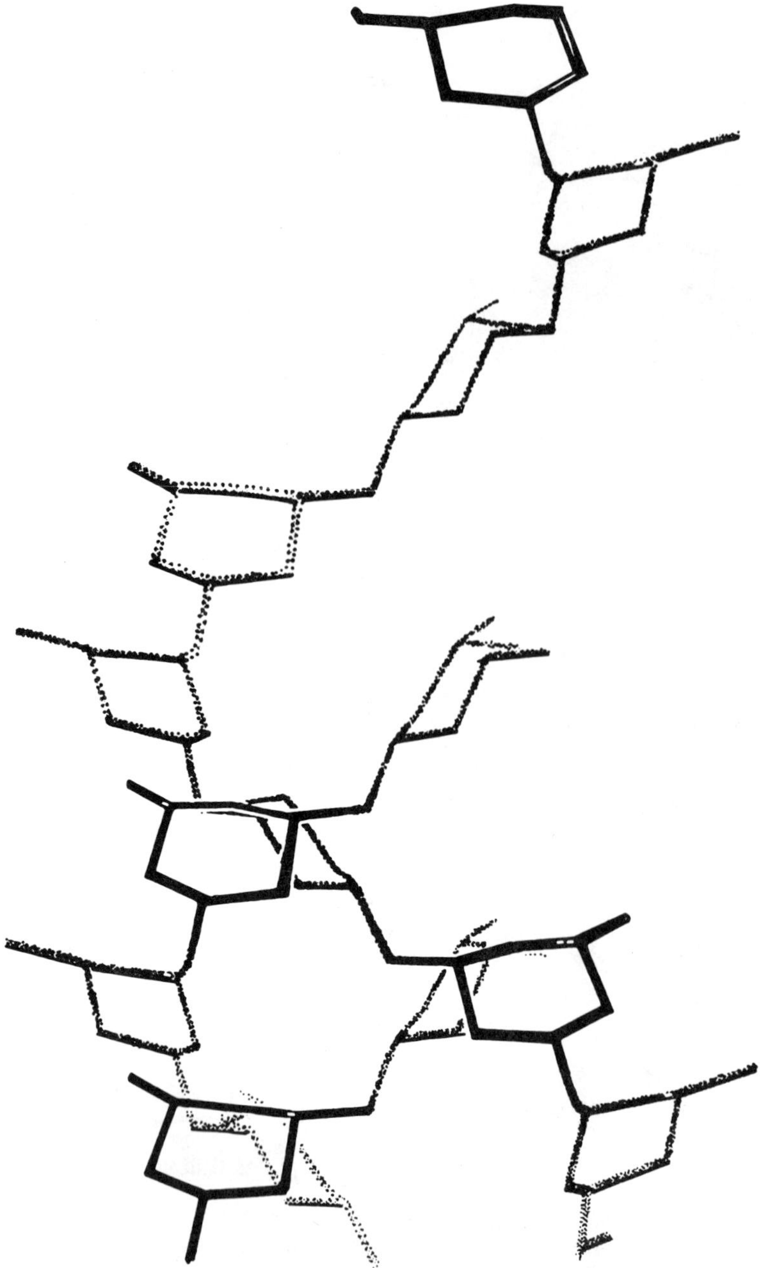

Figure 12. Triple helical arrangement in the crystalline state of β(1 → 3) glucan. Space-filling for the large amplitude helix is achieved by triplex formation.

Such molecules have great potential for uses such as suspending agents, coating, gelation and in secondary oil recovery. More important, their production is based on bacterial and fungal systems which offer great future potential for improvements through genetic manipulation. Furthermore, they are produced from waste biomatter hence fit into the renewable materials cycle.

CONCLUSIONS

The "Biomass and Biotechnology" phase of cellulose science is just beginning. The approach now being used is still mostly soft technology, i.e., pollution of paper mills is being circumvented by preventive means and the viscose process is replaced by new organic solvents. The hard technology has yet to make its full impact felt. The forest refinery will probably not be a reality for another generation since much of the processing will be microbiological. Low energy consumption and high efficiency, thanks to genetic engineering, will provide the route to petrochemical substitution options based on biomass.

The potential of the new biomass technology is clearly seen in food technology where high fructose corn syrup based on immobilized amylases and isomerases has brought about a major economical transformation. Our dependance on the world price of sucrose is a thing of the past.

Conservation philosophy is always the forerunner of new technology as industry tries to stretch the lifetime of its investment. Nowhere is this more evident than in the pulp and paper industry where whole tree pulping is leading to 90% pulp yields for papers which formerly used 45% yields. Can biopulping concepts be far away?

No area of cellulose technology is more ripe for improved productivity than the forest industry. Compared with the green revolution in agriculture, silviculture for commerce is primitive. It is hardly more advanced than the 19th century U.S. farmer. The industry, at best, is at the tree farmer stage when it should be aiming for yield increases which anticipate the biomass needs of the next millennium. The urgency of the situation may be judged from the forest utilization data of Table II. Because trees are biological systems, the application of conventional and less conventional genetics promises a 3- to 4-fold increase in productivity.(38) The steps to achieve this are an essential part of the biomass revolution.

TABLE II

Distribution and Utilization
of World Forests in Millions of Hectares (40)
(1 Hectare = 2.47 Acres)

Location	Forest Area	Accessible Forests	Utilized Forests
North America	710	312	220
Central America	271	329	83
South America	830		
Africa	700	284	108
Europe	138	133	130
USSR	738	425	350
Asia	500	311	232
Australia	92	20	17

Ever since the beginning of polymer science, and long before, cellulose has been part of man's innovation challenge. Broadening the commercial polymer concept to include carbohydrate polymers based on biotechnology is providing a multi-language "translation of scientific findings into useful goods and services" which is Milton Harris' definition of technical innovation.

LITERATURE CITED

1. Harris, M.; Chem. Eng. News, 1980, March 31, 38-41.
2. Payen, A.; Compt. Rend., 1838, 7, 1052-1056.
3. Freudenberg, K.; and Blomquist, C.; Ber., 1935, 6813, 2070-2082.
4. Mark, H. F.; Chem. Eng. News, 1976, April 6, 176-183.
5. Natta, G.; U.S. Patent No. 3,112,300; also Natta, G.; and Corradini, P.; J. Polymer Sci., 1956, 20, 251.
6. Avery, O. T.; MacLeod, C. M.; and McCarty, M.; J. Exp. Med., 1944, 79, 137.
7. Larm, O.; and Lindberg, B.; Adv. Carbo. Chem. and Biochem., 1976, 33, 295.
8. Johnson, D. L.; U.S. Patent Nos. 3,447,939 and 3,508,941.

9. Chanzy, H.; Dube, M.; and Marchessault, R. H.; J. Polymer Sci., Letter Ed., 1978, 17, 219.
10. Malcolm, E. W.; and Scott, W. E.; "Future Trends in the Paper Industry - III", TAPPI publication, March, 1979.
11. Myerly, R. C.; Nicholson, M. D.; Katzen, R.; and Taylor, J. M.; CHEMTECH, 1981, 11, 186-192.
12. Delong, E. A.; Canadian Patent No. 1,096,374.
13. Marchessault, R. H.; Coulombe, S.; Hanai, T.; and Morikawa, H.; Transactions - Pulp and Paper Magazine of Canada, 1980, 6, TR52-56.
14. Chua, M. G. S.; and Wayman, M.; Can. J. Chem., 1979, 57, 1141.
15. Calvin, M.; Chem. Eng. News, March 20, 1978, 30-36.
16. Lipinsky, E.; Science, 1978, 199, 644.
17. Demain, A. A.; "Annual Reports of Fermentation Processes", Tsao, G. T., Ed.; Academic Press, N.Y., 1980, Vol. 4, pp. 193-208.
18. Private communications: Suomen Sokeri, Helsinki, Finland.
19. Zobel, B. J.; J. Forestry, 1974, 72, 681-685.
20. Marchessault, R. H.; and St. Pierre, J.; "Future Sources of Organic Raw Materials", St. Pierre, L.E., and Brown, G.R., Ed.; Pergamon Press, Toronto, 1980, pp. 613-625.
21. Thomas, R. J.; "Wood Technology: Chemical Aspects", Goldstein, I. S., Ed.; ACS, 1977, pp. 1-23.
22. Jurasek, L.; Dev. in Industrial Mibrobiol., 1979, 10, 177-183.
23. Lepoutre, P.; Hui, S. H.; and Robertson, A. A.; J. Appl. Pol. Sci., 1973, 17, 3143.
24. Andrade, J. D., Ed.; "Hydrogels for Medical Applications", ACS Sympsoium Series, Wasington D.C., 1976, 31.
25. Dean, W. L.; and Ferguson, G. N.; U.S. Patent No. 3,589,364.
26. Lassen, F. O.; U.S. Patent No. 3,997,647.
27. Marchessault, R. H.; Dube, M.; St. Pierre, J.; and Revol, J. F.; "Fibre-Water Interactions in Paper-Making", Transactions of the Symposium Held at Oxford, September, 1977, Fundamental Research Committee, Ed.; the British Paper and Board Industry Federation, 1978, Vol. 2, pp. 795-818.
28. St. Pierre, J.; M.Sc. Thesis, Department of Chemistry, Universite de Montreal, Montreal, Canada, 1978, pp. 68-75.

29. Perez, S.; Chanzy, H.; Maia, E.; and Peguy, A.; "27th IUPAC Symposium on Macromolecules (Strasbourg) - Abstracts", 1981, Vol. II, pp. 910-913.
30. Maia, E.; Peguy, A.; and Perez, S.; Acta Cryst., in press.
31. Chanzy, H.; Nawrot, S.; Peguy, A.; Smith, P.; and Chevalier, J.; J. Polymer Sci., Physics Ed., in press.
32. Franks, N. E.; and Varga, J. K.; U.S. Patent Nos. 4,145,532 and 4,196,282; also McCorsley III, C. C.; and Varga, J. K; U.S. Patent No. 4,142,913.
33. Cottrell, I. W.; "Fungal Polysaccharides", Sandford, P.A., and Matsuda, K., Ed.; ACS Symposium Series, 1980, 126, pp. 251-270.
34. Rees, D. A.; and Scott, W. E.; J. Chem. Soc., 1971, B, 469.
35. Marchessault, R. H.; and Deslandes, Y.; Carbohydrate Polymers, 1981, 1, 31-38.
36. Deslandes, Y.; Marchessault, R. H.; and Sarko, A.; Macromolecules, 1980, 13, 1466-1471.
37. Yanaki, T.; Norisuye, T.; and Fujita, H.; Macromolecules, 1980, 13, 1462-1466.
38. Hall, F. K.; "Proceedings of the Conference on Paper Science and Technology", The Institute of Paper Chemistry, Appleton, Wisconsin, 1980, pp. 15-18.
39. Joseleau, J. P.; "28th IUPAC Congress - Abstracts", Vancouver, Canada, Abstract No. PH52, 1981.
40. Information derived from FAO Reports, U.S. Forest Service data and American Paper Institute statistics.

RECEIVED March 12, 1982.

5

The Changing Face of Textile Science

Arnold M. Sookne

Burlington Industries, Corporate Research and Development, Greensboro, NC 27420

This paper provides a brief survey of textile science of the 1930's, and of the golden age of American textile science which followed for about 25 years. The survey relates particularly to the contributions of Dr. Milton Harris, a pioneer in several fields of textile science.

An analysis is presented of some of the factors now affecting research and development in general and textile research in particular. It appears likely that textile science will move farther and farther from basic studies and closer to engineering and technology. This will follow from the demands for conservation of energy and water, minimizing of pollution, and acceleration of manufacturing pro-cessess. Automation will be accomplished with the help of modern computational, sensing and communi-cating equipment.

Textile Science Fifty Years Ago

As long as 50 years ago, textile science had advanced to a high degree in England and on the continent of Europe. In England, for example, J. B. Speakman was doing sophisticated research on wool fibers and fabrics (1), and Fred Peirce was studying the mechanics of fabric structures, examining the relationships between fabric structure and hand, crease recovery, and other fabric properties (2). Foulds, Marsh and Wood were inventing the famous T. B. L. process for making crease resistant rayon (3), which led ultimately to the durable press revolution throughout the world of textiles. W. T. Astbury in England (4) and Meyer and Mark (5), Hess (6), and others on the continent were studying the structures of fibers by physical and chemical means. And in 1932, Hermann Staudinger published his seminal work on high molecular weight organic compounds (7), which was to have a profound influence on the thinking of fiber, textile, and polymer chemists.

8412-0740-2/82-0083$06.00/0

Textile science in the United States was far less advanced than its European counterpart. There were a few centers such as the Massachusetts Institute of Technology and the National Bureau of Standards where important research work in textile mechanics was being conducted by Professor E. R. Schwartz (M.I.T.) and H. F. Schiefer (N.B.S.). In general, however, American textile research projects tended more toward testing and standardization than basic research. Wallace H. Carothers had initiated a brilliant series of studies of high-molecular-weight condensation polymers in 1929 (8), but it was some years before this work affected American textile science. In the field of cellulose structure, for example, a leading hypothesis, propounded by Wanda K. Farr in 1934, proposed that cotton cellulose was built of small particles that coalesce to form fibrils (9). Farr believed that the particles did not dissolve in cuprammonium solution, generally considered to be a solvent for cellulose.

The controversy concerning cellulose particles raged for a number of years, and was resolved in part by the demonstration by Hock and Harris that the Farr particles observed in cuprammonium solution were either artifacts produced by the solvent (Figure 1) or non-cellulosic impurities that had not been removed from the cotton fiber (10).

The controversy was more completely resolved shortly thereafter by R. Bolling Barnes and C. J. Burton, who were applying the electron microscope to the study of fibers. Barnes was able to demonstrate that the apparent Farr particles observed in swollen cotton in the light microscope were actually intersections of fibrils when observed in the electron microscope, as shown in Figure 2 (11).

A general idea of the nature of textile research being conducted in the early to middle 1930's is given by Figure 3, which shows the table of contents of the journal Textile Research in February, 1935. This more-or-less typical listing contains Hardy's significant study of methods for measuring fiber fineness, but overall the subject matter was not too exciting.

The Golden Age of Textile Science

General Comments. American textile science in the early 1930's was due for a renaissance. Synthetic fibers were just beginning to appear, and the manufacturers of such synthetics, as well as of viscose rayon and acetate, were interested in learning more about their products and the textiles which could be made from them. A backlog of research developed in Europe was available for assimilation and application. Finally, there was a period of enormous stimulation by the Armed Services before and during World War II. In this, Dr. Steven J. Kennedy of the U. S. Army played a leading role. Thus, there was a period of flowering and growth that started about 1935 or 1940 and lasted for about 25 years. Among the major contributors to this period

A

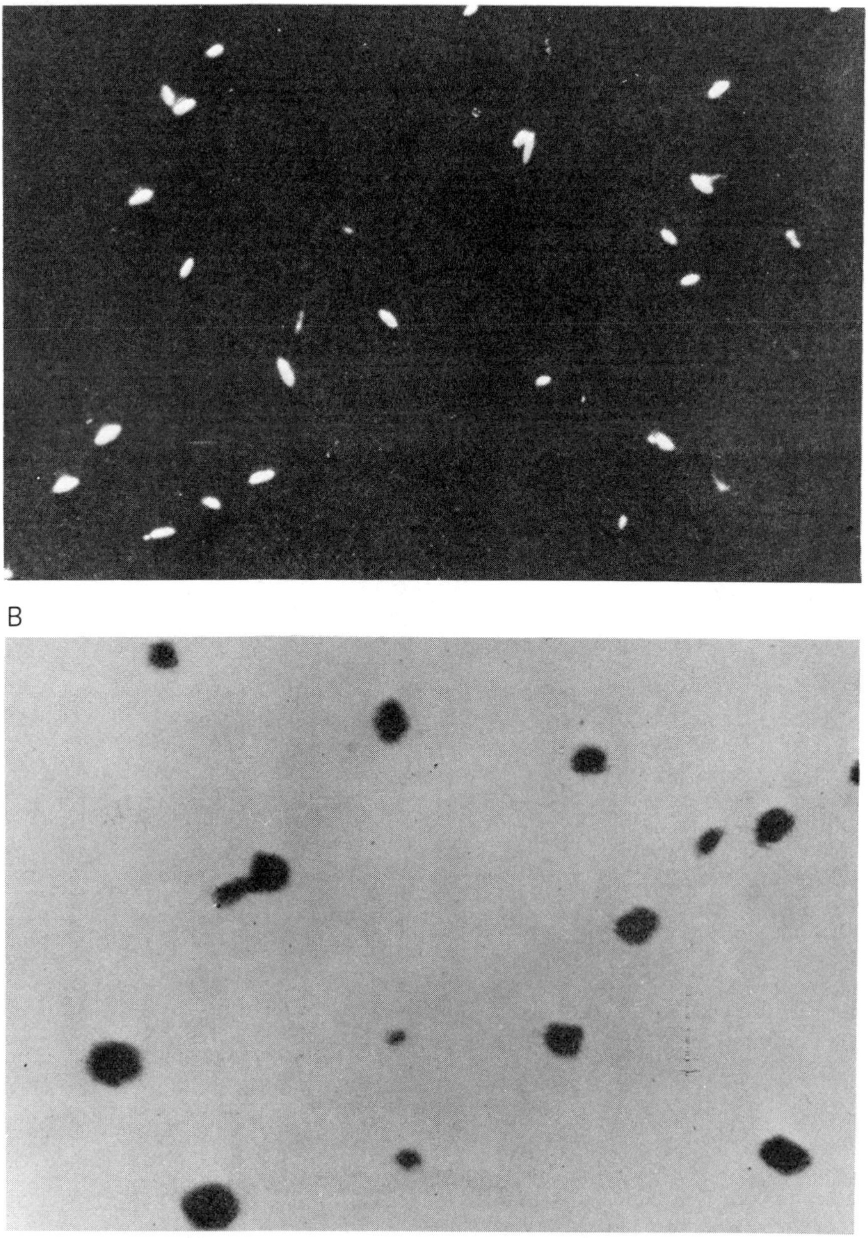

B

Figure 1. Particles formed in cuprammonium hydroxide solution in the absence of cellulose (9). A: Particles as viewed between crossed nicols. B: Particles that appear after addition of iodine and sulfuric acid. Magnification, 1500 ×.

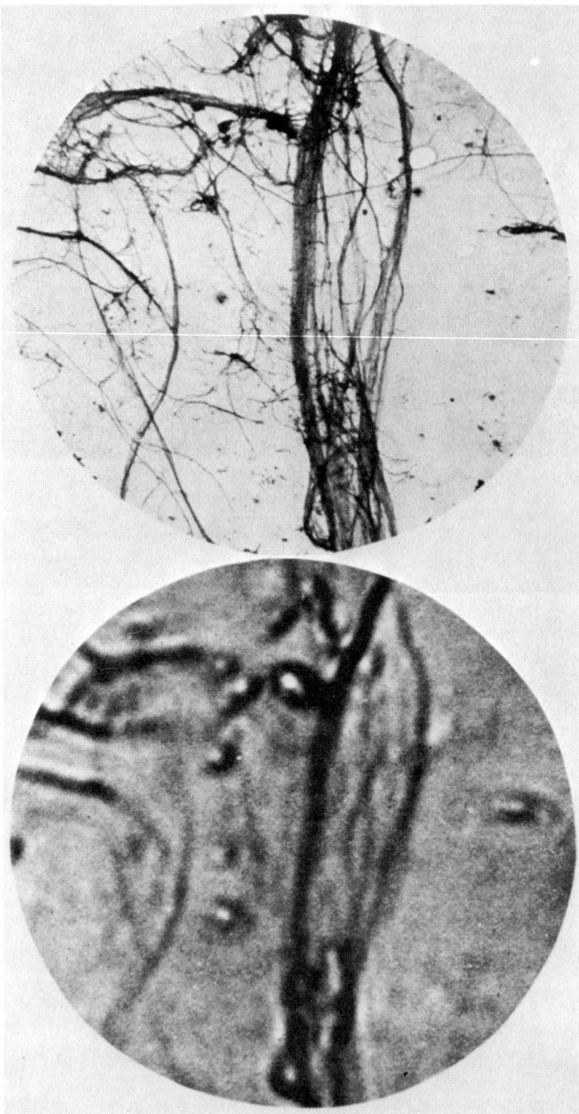

Figure 2. *Comparison of photomicrograph (left) and electron micrograph (right) of the same field of cellulose (10).*

TEXTILE RESEARCH

Publication of

United States Institute for Textile Research, Inc.

CHARLES H. CLARK, *Managing Editor,*

Publication Office: Prince and Lemon Sts., Lancaster, Pa.
Editorial and General Offices: 65 Franklin St., Boston, Mass.

Board of Editors	Publication Committee
DR. W. E. EMLEY	WILLIAM W. BUFFUM
PROF. L. A. OLNEY	ERNEST N. HOOD
PROF. E. R. SCHWARZ, *Chairman*	CHARLES H. CLARK, *Chairman*

Vol. V — February, 1935 — No. 4

CONTENTS

Entered as second-class matter, November 17, 1932, at the Post Office at Lancaster, Pa., under the Act of Congress of March 3, 1879.

Issued monthly at Prince and Lemon Streets, Lancaster, Pa., for United States Institute for Textile Research, Inc., Boston, Mass..
Made in United States of America.

Subscription price $5.00 a year. Single copies .50.

Figure 3. Table of contents of the Textile Research *for February 1935.*

of growth were the Textile Foundation (later the Textile Research Institute), universities such as Massachusetts Institute of Technology, Fabric Research Laboratories (particularly in the field of textile mechanics), and Harris Research Laboratories (particularly in the field of textile chemistry).

Contributions of Milton Harris and His School

Dr. Harris' first position after his employment as a chemist by the Cheney Silk Mills was at the National Bureau of Standards, where he started a research associateship for the American Association of Textile Chemists and Colorists in 1930. A few years later, the group continued under the sponsorship of the Textile Foundation, which later became the Textile Research Institute. A photograph taken in 1938 of the group of ten scientists (in the presence of thirteen committee members) is shown in Figure 4. Dr. Harris is the fourth person on the right, immediately behind his secretary. From this group, there flowed a stream of papers which threw light on fiber structure, functional properties of fabrics, and other subjects of fundamental interest to textile chemistry. Their work during this period principally concerned the natural fibers, since the only manmade fibers at the time were rayon and acetate.

In 1945, the Textile Research Institute moved to Princeton, New Jersey, and Dr. Harris established Harris Research Laboratories in Washington, D. C. Figure 5 shows a picture of part of the laboratory staff of 33, taken about 1948. Many of the people in this picture are here today, several of them are participating in this program.

Basic Contributions. The early work of Milton Harris and his colleagues led to numerous advances in fiber and textile chemistry, particularly for wool and cotton. It's beyond the scope of this paper to list these advances in detail or even to tabulate the references, but some idea of their scope is given by the following range of subjects:

1. Nature of the interaction of wool and cotton with acids and bases (with Jacinto Steinhardt).
2. Microscopic structure of natural fibers (with Charles Hock).
3. Frictional properties and felting mechanisms of wool (with Daniel Frishman).
4. Comfort properties of textiles (with Lyman Fourt and Norman Hollies).
5. Nature of the surface of textile fabrics (with Herman Bogaty and Norman Hollies).
6. Surface chemistry and physics of fibers, yarns and fabrics (with Anthony Schwartz).
7. Polymolecularity and mechanical properties of cellulose acetate (with Arnold Sookne).
8. Photochemistry of wool and silk (with Henry Rutherford).

Figure 4. Staff of Textile Foundation Research Associateship at National Bureau of Standards (with members of advisory committee) in 1938. Dr. Harris is the fourth person on the right, partly behind his secretary.

Figure 5. Part of the staff of Harris Research Laboratories in 1948. Dr. Harris is the fourth person from the right in the front row.

9. Alteration of the molecular structure of wool (with Walton Geiger and Alfred Brown).

This last is probably the work for which Dr. Harris is best known (12, 13). Figure 6 presents a simplified version of the chemical route by which the alkali-sensitive disulfide linkages in wool can be replaced by bis-thioether linkages of greatly enhanced stability. It is probable that this work provided important stimulation to other textile scientists in the general field of chemical modification of fibers.

Practical Developments. The basic contributions mentioned above led to a number of notable practical textile developments, a few of which are listed below:

1. A practical, low-cost process for producing shrink-resistant wool (with Daniel Frishman). This process was used by the United States Army to produce many millions of pounds of shrink-resistant wool socks, sweaters and mufflers.
2. Comfortable clothing for the United States Army (with Lyman Fourt and Norman Hollies).
3. Improved water-repellent fabrics for Army uniforms (with Francis Minor and Arnold Sookne).
4. Practical processes for stripping color from wool and bleaching cashmere and other animal fibers (with Alfred Brown).

Other practical developments that emerged from the Harris group dealt with the non-textile fibers, human hair and beard hair. These included the following:

Heat-generating shaving cream, in which an exothermic reaction was used to generate heat as foam was expelled from a can.

Improved hair waving products.

Improved razors through polymer coatings on the edges of blades.

The conventional methods of chemistry and physics are often not directly applicable to textile science, so that the development of reliable methods is often an essential part of textile research. Among the numerous developments of Dr. Harris in this field, the simplest and oldest is the alkali solubility test for determining the state of oxidation of wool. This humble test, developed in 1936 by Harris and Smith (14), is still in use today as a means for determining damage to wool. Indeed, it is still an official test of the American Society for Testing and Materials, which may qualify it for a place in the Guinness Book of Records.

During the 25-year period that has been characterized as The Golden Age of Textile Science, there were equivalent flows of ideas and results from other centers of textile research, both in the United States and abroad. It is beyond the scope of this presentation to summarize these results, but among American laboratories, special mention should be made of Fabric Research

Laboratories, headed by Walter Hamburger, which was extremely productive for many years in the field of fiber and textile mechanics. Their results added greatly to our understanding of creep in textile structures, abrasion resistance, and the bending and recovery properties of fabrics. Special mention should be made as well of the Textile Research Institute, which continued in Princeton with a reorganized staff, and which has continued to do outstanding research and development work in fiber and textile chemistry and mechanics, flammability, and other fields.

Textile Science of Today and Tomorrow

General Comments. The uncertainties involved in predicting the future are well known, and have humbled even great persons who have made predictions in their own field of expertise. (Thomas Edison is reported to have made this comment in 1889: "I have always consistently opposed high-tension and alternating systems of electric lighting. . .not only on account of danger, but because of their general unreliability and unsuitability for any general system of distribution.") Who can predict, for example, whether the conservation ethic will become a dominant force in our society? How long will the trend toward the "natural" last in clothing and textiles? While questions such as these cannot be answered with any certainty, it is relatively simple to list the obvious current factors that will almost certainly affect textile research in the next decade and to estimate qualitatively their probable effects. In one group are the adverse factors of increasing costs for energy, environmental controls, water, labor, and capital. In a second group are the beneficial factors of decreasing costs of computation, sensing, analysis, and control. In a third group are societal factors, such as the continuing trend toward the "natural" (or sometimes the unnatural or even the occult), and changes in indoor living conditions stemming from the energy crisis. The latter will, of course, mean colder rooms in the wintertime and warmer rooms in the summertime.

Adverse Factors. Let's examine the adverse factors in a little more detail. As shown by Figure 7, which depicts the relative costs of coal and No. 2 heating oil, there has been more than a sevenfold increase for oil between 1972 and 1980, while the cost of coal has increased about a third as much. The projections that are shown for 1980 through 1990 suggest that the cost of oil will continue to rise at a very rapid rate, while coal will increase in cost more moderately, at least by current standards of what is moderate. It seems clear from these data that process improvements which reduce energy consumption will continue to have high rewards. Project work in energy conservation will almost inevitably consume manpower that would have gone into longer-range R&D.

A second adverse factor is the cost of meeting health and

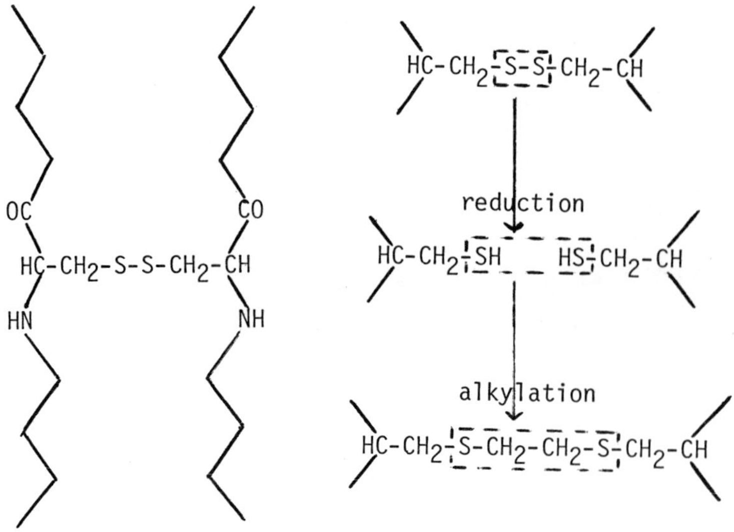

Figure 6. Left: Schematic representation of disulfide link between polypeptide chains in wool. Right: Reduction of disulfide linkage (top) to sulfhydryl groups (center) and alkylation to form stable bis(thioether) linkage (bottom).

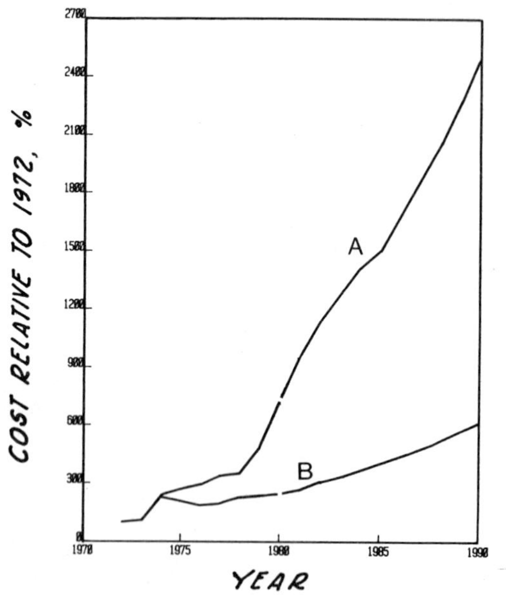

Figure 7. Relative past and projected costs of No. 2 fuel oil (A) and coal (B).

safety requirements for our industry, for which Figure 8 attempts to provide a qualitative estimate. These costs started to increase toward the end of the 1960's, and have risen quite rapidly in the last five years. The experts differ with respect to the probable future trends, the current conventional view seems to be that there will be a leveling off of the increase in the 1980's, particularly in view of the current political climate. Whether costs increase steeply or at a moderate rate, however, government-mandated work will continue to absorb a significant fraction of the manpower that otherwise could be applied to textile science.

A third adverse factor is the increasing cost and scarcity of water. It is difficult to obtain an average cost for water in the United States, since it depends greatly on location, the availability of wells, and related factors. However, about half of the cost of water is attributable to the cost of energy for pumping and purification. It therefore seems reasonable to expect significant increases in cost over the next decade, as shown schematically by Figure 9. This curve represents a guess, rather than a prediction. Nevertheless, water conservation must be rated close to energy conservation as a factor that will swing textile science toward technology and engineering in the coming decade.

Figure 10 shows the average hourly gross earnings of textile production workers, taken from data of the American Textile Manufacturers Institute. Wages have increased roughly two-fold over the last ten years at a more or less linear rate. If inflation continues at its present pace, we can expect an escalation of this rate. This factor mandates increased engineering activity in automation and increased speeds of production.

A final adverse factor that merits consideration is the cost of borrowing money, as shown by the prime rate. Data for this factor are readily available, but the fluctuations of the prime rate have been so dramatic within the last couple of years that no sensible pattern emerges from a graph or table. However, the general comment can be made that the prime rate has reached historically high values during recent years, and that it tends to remain high in periods of high inflation. Since high inflation appears to be with us for the foreseeable future, it seems probable that the cost of borrowing money to purchase new machinery will remain high. Certain types of machines, such as high-speed looms, must be purchased by any progressive textile company which expects to remain in business, but this is not true for all machines. In the case of tenter frames, scouring equipment, and dyeing machines, the alternative of optimizing the operation of the machine through the use of modern sensors and computer technology offers an attractive and often effective alternative. Such changes require the application of known technology and engineering, rather than new textile science.

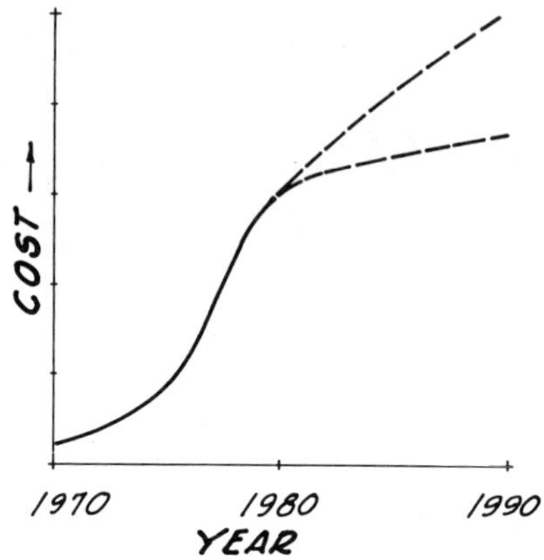

Figure 8. *Schematic representation of costs of meeting government health and safety requirements.*

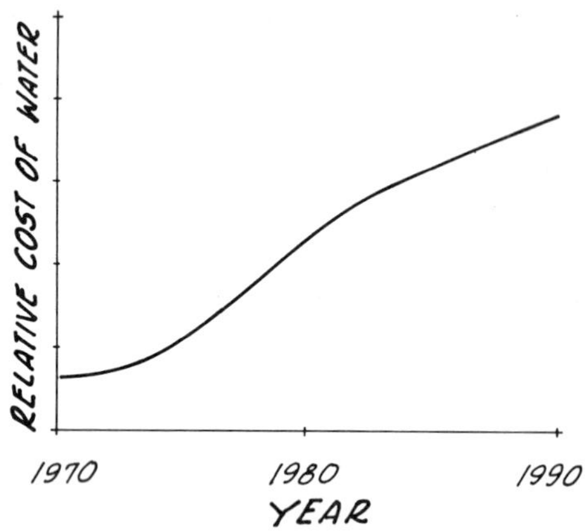

Figure 9. *Schematic representation of the changing cost of water.*

Beneficial Factors. Let's turn to some of the more positive factors which will influence R&D in the 1980's. Figure 11 is intended to provide one measure of the cost of computation by showing the price of computer memory as a function of time from 1964 to the present. There has been a dramatic decrease in these costs from about $2.00 per byte to about 1¢ per unit in 1979. Furthermore, the curve continues to drop remarkably, and the ability of computers and microprocessors to perform calculations rapidly and cheaply appears to grow almost without limit. These remarkable developments have already appeared not only in our calculating machines, but in our automobiles, television sets, and electronic devices of all types. They can hardly fail to have an important and beneficial effect on textile manufacturing through improved engineering and technology.

There are several additional favorable factors for change which can be expected to help textile R&D in the next ten years. Improvements in computation make possible mathematical modeling of textile processes and machines, such as scouring, dyeing, or tenter framing. In essence, this procedure creates a series of equations that can simulate a tenter frame, for example, and can optimize conditions for any given fabric. The mathematical model can even exercise interactive control; that is, it can maintain optimum running conditions as the fabric passes through the machine.

In the field of communication, new developments with optical fibers, or light pipes, are lowering the cost of carrying information from a machine to the computer. The tieing of the computer to the telephone and to communications satellites also leads to more rapid, lower-cost transmission of information.

Other favorable factors relate to improved devices for sensing temperature, for example by means of temperature-sensitive phosphors made from such esoteric compounds as lithium tantalate or gallium arsenide. Such sensors, coupled to optical fibers, permit the precise sensing of temperature in unusual and hostile environments. Sophisticated infrared analytical devices, coupled to a computer, are capable of "learning" to analyze for polyvinyl alcohol in the presence of starch, or to make other difficult analytical separations. New, rugged transducers permit the monitoring of pressure or temperature in difficult environments. Many of the new devices for sensing, analyzing, and communicating results from work done by NASA, as well as by the computer and instrument companies.

Effects on Future R&D. How should these factors for change, both favorable and unfavorable, affect our R&D programs in the 1980's? In the field of dyeing we should concentrate our efforts on continued reduction of liquor ratios and wet pickups, through the use of foams, effective padding, jet dot printing, etc. The cost of water nationally is given at about $1.00 per thousand gallons, according to the Environmental Protection Agency, although it is more like 60¢ per thousand gallons in a

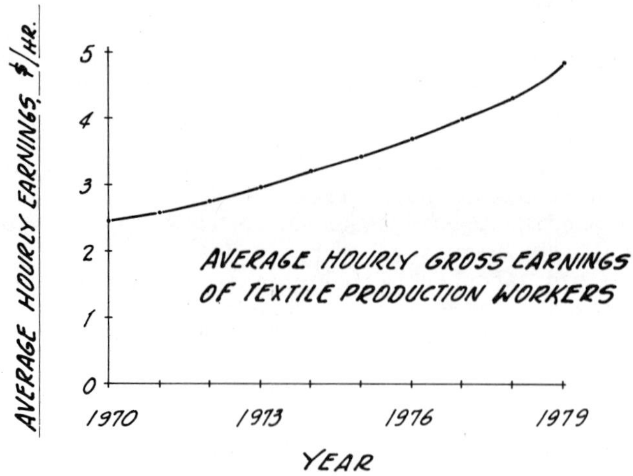

Figure 10. Cost of labor, 1970–1979, for textile production workers.

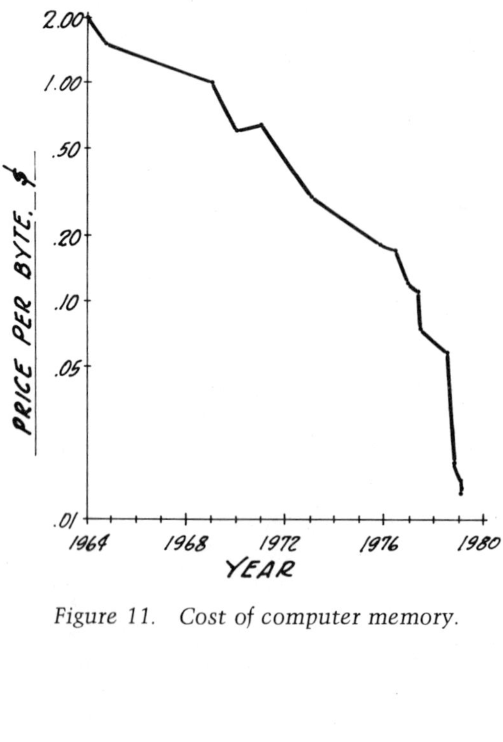

Figure 11. Cost of computer memory.

medium sized city like Greensboro, North Carolina. A new desalination plant scheduled to operate in Yuma, Arizona by 1985 will produce 1,000 gallons of partially desalinated water suitable for agriculture for about 81¢. It seems reasonable, therefore, to expect that the reuse and recycling of water will become attractive in many locations during the next decade. This work, too, will absorb funds and manpower that might have gone into longer-range R&D.

In the field of dyes, we can expect that fewer manufacturers will be making fewer dyes, largely because of the cost of environmental controls and the need to remove dyes which contain hazardous components. On the other hand, the widespread use of the computer for choosing dyes and formulas can permit operation with a smaller number of dyes. With appropriate engineering effort, the computer can be expected to be used for optimizing and even controlling the dyeing process in both batch and continuous operations. And finally, we can expect that the computer and spectrophotometer will work together for the sorting of rolls of fabric to minimize seconds in garments. The 1980's may see the spread of no-add dyeing, and the disappearance of the "dyer" as such; he may be supplanted by a machine operator. Other changes may involve the more widespread use of the ink jet, computer controlled, for not only printing but the dyeing of solid colors. Most of these changes can be made by applying existing technology, or by developing it to a reasonable degree.

What are some of the corresponding goals that are appropriate for finishing in the 1980's? Certainly the reduction of the use of water and therefore energy in scouring must be high on this list. So must the reduction of wet pickups, either by foam or effective padding. The reuse and recycling of water and the recovery of polyvinyl alcohol and other chemicals from the water can be expected to grow with appropriate R&D effort. Few new chemical finishes should be expected, because of stringent government regulations on such new developments. And it is probable that any finish containing formaldehyde is likely to suffer. Since it has been demonstrated that formaldehyde can, in fact, cause cancer in rats by inhalation, the use of formaldehyde is very likely to be reduced, or perhaps even banned. Formaldehyde-free finishes, or fabrics that require no chemical stabilization, may be required.

Other changes in finishing procedures will probably include more stringent environmental controls with respect to effluents, and even the solid waste in our lagoons. It appears almost certain that the computer will be in widespread use for the control of scouring and finishing, for automatic inspection, and for modeling of the tenter frame.

Effects on Textile Technology. In the field of textile technology - the spinning of yarns and formation of fabrics - many of the above-mentioned factors will also operate, especially the ever-present factor of increasing energy costs. In this field, progress will probably be provided primarily by the manufacturers of textile machinery. In the last few years, this industry has provided us with carding machines of enormously increased speed, as well as open-end spinning machines that depart radically from the classic ring-spinning frames and are capable of spinning heavy yarns much more economically. The classic loom which utilized a shuttle to carry the filling yarn across the warp is on the way out, and is being replaced by new looms that utilize small carriers, water drops, or air jets to propel the filling across the warp. The new looms are capable of much higher speeds and greater widths, so that the rate of production has been increased about eight-fold in the last 40 years. While these new machines are often produced at the demand of or in cooperation with the textile industry, they represent engineering developments of the machinery industry, and once purchased, they have an important effect on the nature of the fabrics produced by the textile manufacturer. To a considerable degree, leadership in textile technology has moved from the textile scientist in the university, institute, or industry to the machine manufacturer. A further aspect of this loss of leadership in textile technology concerns the development of new filament yarns by the fiber manufacturer. These new filament yarns, which possess improved esthetic characteristics and bear a close resemblance to staple yarns, are being developed by the fiber manufacturers on the basis of careful long-range research. Once provided, the new yarns, if successful, will also have an important effect on the characteristics of the fabrics produced by the textile manufacturer.

Effects of Societal Factors. A final set of factors that can be expected to affect textile science in the coming years concerns societal changes that can be expected as a result of the energy crisis, as well as the continuing trend toward the "natural". With respect to the former, the effects are predictable; we can expect that the cooler indoor climate in the winter will lead to an increased demand for warm clothing with a natural feel, made from fabrics such as corduroys, velvets, napped fabrics, etc. Warmth depends primarily on the thickness of the trapped air layer rather than fiber composition, so that the thicker fabrics with considerable surface finish should become more popular, whether they are made of synthetic or natural fibers. Conversely, one can expect that the warmer indoor environment in the summer will lead to the demand for cool permeable clothing with a natural feel. Such fabrics will be relatively thin and porous, but again it should be noted that the fiber composition has little effect on water vapor permeability. For most apparel fabrics, water vapor permability is a function primarily of fabric thickness and structure, and not fiber composition. Here textile science can play an important role in the development of new, more functional fabrics and garments.

With respect to the second factor, the trend toward the

"natural", predictions are less clear. Most of the experts in predicting social trends apparently believe that the conservation ethic will grow, and the trend towards the "natural" may well be diverted to natural-appearing and natural-feeling textiles and garments. Interesting and important consumer preference studies are currently being performed in this area.

In summary, the major trends discussed above include the following: A rigorous optimization of the use of energy and minimization of the use of water is essential. Recycling and reuse of water and some chemicals can be confidently predicted. Greatly expanded use of microprocessors and computers for control, optimization, inspection, etc., is almost inevitable. And with respect to new fabrics, which will require new finishes, we can expect the growth of thick, warm fabrics for the wintertime, and thin, porous fabrics for summer use. These, then, are major changes which can be predicted, and which will be forced upon us for the most part. They represent opportunities for the application of engineering science and technology, rather than longer-range R&D.

Epilogue

Most of us who worked with Dr. Harris never stopped to ask ourselves why there was such a strong feeling of cooperation in the group, why we worked hard and were interested in our work, or indeed, why the group was successful. We never stopped to ask why we were constantly visited by the world's greatest textile scientists, not only from this country, but from England, Europe, Australia, and Asia. It is perhaps true that the work ethic was more popular at that time, particularly during depression days. However, the overriding factor was the brilliant, effective, and warmly human leadership provided by the man whose 75th birthday we celebrate today.

Literature Cited

1. Speakman, J. B. J. Soc. Chem. Ind. 48, 321T (1929).
2. Peirce, F. T. J. Textile Institute 21, T377 (1930).
3. Foulds, R. P.; Marsh, J. T.; Wood, F. C. Brit. Pats. 291,473 and 291,474 (1926).
4. Astbury, W. T.; Street, A. Philos. Trans. (A) 230, 75 (1931).
5. Myer, K. H.; Mark, H., "Aufbau der hochpolymeren organischen Naturstoffe"; Leipzig, 1930.
6. Hess, K. "Die Chemie der Zellulose und ihrer Begleiter"; Akadem. Verlagsgesellschaft: Leipzig, 1928.
7. Staudinger, H. "Die hochmolekularen organischen Verbindungen" J. Springer: Berlin, 1932.
8. Carothers, W. H. J. Am. Chem. Soc. 51, 2548 (1929).
9. Farr, W. K.; Eckerson, S. H. Contribs. Boyce Thompson Inst. 6, 189 (1934).

10. Hock, C. W.; Harris, M. J. Research Nat'l Bur. Standards 24, 743 (1940).
11. Barnes, R. B.; Burton, C. J. Ind. Eng. Chem. 35, 120 (1943).
12. Harris, M. Textile Research 5, 19 (1941).
13. Geiger, W. B.; Kobayashi, F. F.; Harris, M. J. Research Nat'l Bur. Standards 29, 381 (1942).
14. Harris, M.; Smith, A. L. J. Research Nat'l Bur. Standards 17, 577 (1936).

RECEIVED May 1, 1982.

6

American Science Policy in Practice

W. O. Baker

Bell Laboratories, Murray Hill, NJ 07974

Successful titles such as newspaper headlines, we are told, should have bold, even if frequently false, assertions. A little of that tactic may occur in our title implication that national science policy ever meets practice. Indeed, policy of any kind is often not very well designed for practice. But in the case of government, especially large ones like Federal government, there is frequently ingenious and diligent endeavor to avoid testing policy by practice. For policies in Washington obviously soon go into politics. One is likely to find a scenario of a politician having to face whether his policies work into real practices. This test is just too dreadful to comtemplate, to say nothing of experiencing.

But now this whole situation is ruthlessly intensified when the policies are about science and technology, or research and development, and when practice is what makes and keeps air clean, or causes cancer or generates electricity, or launches spaceships or guides rockets for free world security, or launching of interplanetary probes; or what actually happens to employment and economic policy vis ā vis the upsurging technology of automation, wherein the bit hits the man. The trouble with science and technology practice is that you can almost always tell whether it works. Planes fly or fall, bridges span or bend, pills cure or fail, so policy gets tested in the most brutal and visible ways, which policymakers abhor and elegantly avoid.

8412-0740-2/82-0101$6.00/0

The situation thus favors a gross separation between policy and practice -- an action gap, something like the mythical technology gap with which we explained the backwardness of the Japanese and European economies a few years ago! (Such a gap they are increasingly filling with favorable exports of automobiles, consumer electronics, machine tools, and Deutschmarks.)

But back to science and technology, in government and industry, education and the economy. To assure appropriate detachment of policy from practice in science and engineering requires the genius of making things work in practice so well that nobody will much notice that none of them is even supposed to have a policy associated with it! Whereas there have been a multitude of political/social/economic ideas in mid-Century America that would curl your hair, only Milton Harris was able to achieve a technical practice which did it without perturbing the cerebral policy-shaping of the cortex down underneath. Now that's not a bad model for the White House process, where indeed it is particularly important to assure the independence of most of the activities from any cerebral participation anyway, especially if the higher cortex function is in any way implied.

Now this is not to say at all that thinking of an even higher order is also uninvolved in practice -- quite the opposite. It is that science and technology in free enterprise are tested and evaluated steadily and visibly. For instance, in our particular company, the majority of the Western Electric's $12 billion or so output for 1980 was of products simply nonexisting five to eight years ago. They came out of the Laboratories and functioned, or failed to emerge and function, in a form related to what we call employment frequencies (or is it wavelengths) of those responsible for the Laboratories operation and program! Further, "survival" or "span of employment of R&D directors" criteria could encumber the formation of policy, even science policy, and that's why it's nice not to have the two domains, of policy and practice, within too easy reach, or it seems crude to say, within striking distance of each other. The crucial thing, the important thing, the one so appropriate for this happy gathering, is the realization of when they are close enough to interact, and what to do about it then.

So in principle the political positions of research and development in government, and the practical positions in industry, while often happily bridged by the intellectual and pedagogic positions in education, nevertheless have basis for deep division. The interesting matter then is how, if ever, are they brought together? Is it only in America, or principally so, that this long, and tireless search for linkages and commonality seems fruitful? In so sophisticated a parliamentary system as the United Kingdom, we've seen the government's adventures in the National Research and Development Corporation, in the nationalization of steel, coal, transport, some autos, etc., to be unpromising or even abandoned. In France, the CNRS is clearly and efficiently a government venture. Everyone knows that Japan Incorporated is the senior partner of what other incorporations there may be, while in Germany, "Das System" is so regular that even order-disorder phenomena have reduced entropy change! Yet here we are in the USA, with a polymorphic structure of government science and technology policy, linked by innumerable Congressional committees and Executive Branch agencies to political constituencies. These include health care, social security, national defense, agriculture, and an occasional foray, mostly defense-stimulated, into space and nuclear energy. This pluralism is enough so that, of approximately $67 billions of national investment in research and development expected for 1981 (of which about 70.4 percent of $67.4B is performed in industry), about 47.6 percent ($32.0B) of the cost of the total endeavor is actually funded by Government programs. These in turn were initially based on political actions and goals. But the exciting circumstance is that these activities are all tangled and intertwined with an essential and invaluable basis of academic research (which, while Federally funded, is for reasons that we shall emphasize presently, still fortunately not Federally dominated or prescribed). The whole mixture achieves remarkable independence, variety, and creativity, in harmony with the free market and economically-oriented R&D outputs.

This unsurpassed pluralism has immense values in preventing narrowness, stimulating interplay of nominally separated intellects, and altogether speeding up progress. It tends to lack, of course, the intensity and hyper-efficiency of the Japanese automobile quality assurance, technical program, or integrated circuitry cartel. Japan's 1981 industrial R&D spending of 21.6 percent larger than 1980 will quite probably outpace the efforts of some of our competing industries. But overall, in dynamic new fields such as the biosciences and technologies, materials science and engineering, certain energy programs and photonics, as well as the more subtle science and technology of software, we do well to stick with our multiple sources/multiple funding, and peculiar policy/practice hybrids. This is how the total domain of modern electronics, telecommunications and computers came about.

The consequences and formative facts of this polymorphism are intriguing. Above all, they lead to a mixture of citizen and official influcence, and even control, which we believe altogether fits the vast potentials and proven contributions of science and engineering in a democratic society. Beyond these ideological appeals, however, there is another: this situation means we have constructed for ourselves compulsion to have the best and the brightest, working on the practical resolution of the biggest and hardest of scientific and technical matters. Over and over, in these past three decades of Federal science policy, we have seen nations, both allies and opponents, organize for scientific and technical effort in which their leading academic and industrial figures were not significantly tasked or responsible. Officials were always required. I always remember the combination of alarm and outrage that was expressed by some of our most senior charter members of the White House science office when, on the NATO Science Committee (then of high importance for defense of Western freedom,) the Germans, with the best of intentions, assigned a distinguished city planner. He had the bureaucratic official stature; he had negligible science and technology.

Both President Truman and President Eisenhower recognized that even, or perhaps especially, in non-war times the highest skills and talents in the total population should be sought by the Federal government, for worthy inputs. The magic of this situation is obvious in many forms. We should steadily remind ourselves that vital among them is that while constructing policy or aiding in the formulation of policy positions -- for an intercontinental missile, a treatment for cancer, the genetic engineering of soybeans, education and a public library system of nationwide capacity -- these independent and unofficial recruits will have just come from a laboratory, or study, or computer. There they are facing the realities of nature. Thus they will put ideology in its place, as an official often cannot afford to do. Isn't it some such system that our population is seeking even now, in the Federal government's part in economics (including money printing), regulation, criminal justice, and other vital issues?

Anyway, whatever the answer to that, the evidence seems to be that we should guard the way science and engineering are worked out through this pluralism; we should preserve it zealously.

That preservation may not be so easy either. For there are certain aspects of this process of informed and brilliant amateurism in government, bonded to shrewd and competent professionalism in science and engineering, that makes the situation look fragile and sometimes funny. But those really aren't bad features either. We have all seen many such episodes and I just point out a couple that stuck particularly in one's mind. The first PSAC Report to President Eisenhower was on the dissemination of scientific and technical information. It proposed a system in which the basic reports and scientific objectives of all the Executive Branch agencies would be coordinated and available in an "information exchange," which in due time would be automated and accessible to all other branches of government. President Eisenhower, who was keenly sympathetic to science and technology in all forms, decided to take this report to a meeting of his Cabinet and to see about making notice of such reports into regular Cabinet items. His description of what happened at that one, when indignation surged at having one's specific bureaucratic findings put in an accessible exchange of literature, was full of the colorful language for which the General was suitably noted. He pointed out in eloquent detail how the Secretary of Agriculture had declined to expose his fertilizer findings to the skeptical scrutiny of the Secretary of HEW. He concluded that no science report would ever again be exposed to Cabinet policy bickering, and so far as I know, up to now none ever has been.

And so we can decide, after the nearly half-century that we have had the joy of sharing with Milton Harris in his professional, industrial, governmental, and very humane career (we could have said 'adult life,' but Milton has never let adulthood overwhelm his finer, happier instincts) that the nation's science and technology policy and practice have also matured. But we take heart, indeed in seeing how well based have been the early ideas of how the central leadership, the White House science office, could strengthen and extend the lively pluralism of American R&D. Because of our unfailing multiple resources, there has been a ceaseless vigor of creativity and application that Milton Harris has represented so steadily and wisely in his particular ventures. The many events of this admirable symposium amply illustrate what we have asserted. The laws of nature are different than the laws of politicians and courts, (or even the laws of society, which are themselves hardly discernible in the politics and judicial actions of today)! There are 500,000 lawyers in our nation seeking ever to adjust our codes. But there are less than 500,000 scientists and only twice that number of research and development people practicing illumination and application of the laws of all of nature. So we are lucky indeed to have, so far, national science policies of diversity and pluralism, assisting in every way we can to forward the wide participation of our precious talent for generating, organizing and applying new knowledge of science and technology.

The future beckons strongly for the proposals we have recently summarized -- concerning advances in photonics, in materials science and engineering, in building all our great base of electronics and communications, of pharmaceuticals, of molecular biology, of automata and robotics. And in addition, we see coming on chances to do software science and engineering that will literally recast our ways of studying, of action, of production, and of lifestyle. Our professional societies like the American Chemical Society already provide a structure for mobilizing and involving our total national community. They accent especially the precious role of the individual in achieving this future.

In the decades up to this birthday time, this diamond anniversary of Milton Harris, we have created many mechanisms for assuring the pluralism and wide scale involvement of American science and technology in government, without the dangerous or dampening imposition of policy on practice. We believe that the message of the American people to President Reagan and to his Administration does carry over into science and education, technology and industry. It confirms pretty well the position we have taken in these prior years of White House science involvement. It is, that the Federal government doesn't always know best. It is rather touching that the present Administration is already finding that it knows less about what is best or even what is worst than might have been assumed. Nevertheless, we were able in past times to gain excellence through special devices such as the creation in the defense community of appropriate connective tissue for government and industry like the Lincoln Lab, Systems Development Corporation, the Aerospace Corporation, the Applied Physics Laboratory, the Draper Laboratories, the National Laboratories such as Los Alamos, Sandia, Oak Ridge and Argonne. These were ways to create a marvelous system for bringing the diverse R&D resources of industry and civilian affairs together on sufficiently neutral ground so that even a paranoid Department of Justice prosecutor could find little cause for action.

Let us turn now to similar constructions for the future, which will indeed enable government, universities, and industry to pool their skills in meeting world challenge and competition. For we know that larger dimensions of science and engineering that are provided by our many lively but small industrial enterprises are going to be essential in dealing with research and development, invention and discovery in the times ahead. Various significant signals already are in view. More than two years ago, stimulated by the invention of the EBES (or electron beam pattern generating machine) for integrated circuitry in our Laboratories, our associates in the industry led by Dr. Lester Hogan of Fairchild raised the question, particularly with the Defense Department and with others of us about some pooling of knowledge and of resources, even before Japan Inc. begun its latest sweep. Currently the proposals of Control Data Corporation, announced by President Robert Price a couple of weeks ago at the Palm Springs meeting of the Computer and Business Equipment Manufacturers Association, renews the question about such a cooperative. (By the way his phrasing gave us one of the more oblique compliments in our history when he said that a joint venture for VLSI in advanced microcircuit development was needed to offset the impending dominance of "Japan Inc., IBM and <u>possibly</u> Bell Labs." Well our intentions and I'm sure those of our associates at IBM is not to offer an alternative threat to American industry that Japan Inc. represents. Further, however, the fact that we are only <u>possibly</u> part of the dominance is revealing, since neither our esteemed contemporaries at IBM nor Japan Inc. have succeeded in producing microprocessor chips effective for 64 kilobit rapid access memory, nor for 32 kilobit microprocessors of sufficiently able performance to meet communication needs, for which we are now in manufacture. Evidently the information level about creativity and innovation, which we regard with concern in government, can stand also a little improvement in industry.) But this is merely representative of the point we are making that pluralism even in errors and noncredibility has its appeal!

But more seriously, you hear also about the developing nations exporting technology to us, with the South Korean electronics firm, the Gold Star Company with which we already have a joint venture in Korea, building a plant in Huntsville, Alabama for color TV. And not only Japanese but other textile exporters are seizing further initiatives in both the technology and commerce of textiles in the U.S. So we should see rather clearly that we must keep on trying to take advantage of the marvelous flexibility of our pluralism, by using it for combinations of knowledge as well as for counter forces and mediating influences in the inhibiting application of policy.

Milton Harris' career is a fine prototype for the ingenuity we must now exercise in the next stages. His forthcoming decades of service, leading of course to a Centennial Symposium after the turn of the Century, will give us an excellent opportunity to carry on the stimulating tensions between policy and practice. The present National Administration will give us a good start, we believe.

RECEIVED June 29, 1982.

7

The Fine Structure of Keratin Fibers

R. D. B. Fraser and T. P. Macrae

CSIRO, Division of Protein Chemistry, Parkville, Victoria 3052, Australia

Basic studies of keratin-containing fibers such as wool and hair aim to provide a basis firstly for understanding their physical and chemical properties and secondly for modifying these properties for particular end uses. Milton Harris was one of the early champions of this view and it is more than forty years since he delivered an address to the Annual Meeting of the American Association of Textile Chemists and Colorists (1) in which he pointed out that in face of competition from synthetic fibers, the possibility existed of improving the properties of natural fibers . . ." which can only come about by a better understanding of the more fundamental nature of the material". This survey of present-day knowledge concerning the fine structure of wool and hair is offered as a tribute to his pioneering efforts in the application of fundamental studies to textiles and to his belief in the value of "the exchange of fundamental data between . . . different fields."

In attempting to relate the physical and chemical properties of wool and hair to their basic structure it is necessary to consider three factors. These are the *form*, or shape, of the fiber, its chemical *composition*, and the molecular and histological *texture*.

The influence of form is well illustrated by recalling the profound effects of surface profile, diameter, and crimp on the textile properties of wool (2). The influence of composition on physical properties is also well documented, and the compositions of fibers from different animals and from different parts of the same animal show variations which are undoubtedly function-related (3,4,5). In addition to the organic constituents wool and hair fibers contain sorbed water, which must be included in the concept of composition, and which plays an important role in determining mechanical properties (6,7).

8412-0740-2/82-0109$8.25/0

Whilst many properties of wool and hair fibers can be related in a general way to their form and composition, others such as mechanical anisotropy can only be understood in terms of the texture, or spatial organization, of the constituent parts. In the present survey, the topics of cellular texture and composition will only be covered briefly and the main body of the survey will be devoted to the molecular composition and texture of the keratin deposits. Comprehensive accounts of the cellular texture are readily available (8-11).

Cellular Texture and Composition

The principal constituent of wool and hair is a proteinaceous substance termed keratin which is synthesized and laid down intracellularly in specialized cells termed keratinocytes during their passage from the germinal layer at the base of the epidermal follicles to the skin surface. In addition to keratin deposits most hairs contain pigment granules, which are synthesized in cells termed melanocytes. The melanocytes remain in the vicinity of the germinal layer and the pigment granules concentrate in dentritic processes. The transfer of granules from these dentrites to the keratinocytes is believed to involve phagocytosis by the recipient cell (12,13).

In the final stages of fibrogenesis the keratin deposits are stabilized by the formation of disulfide linkages, the regions between cells are extensively modified, and the cells die. The principal histological components of a fully formed hair are an outer layer, or cuticle, of flattened cells, a cortex composed of axially elongated cells, and a central medulla composed of cells which are largely empty.

The Cuticle. During their passage through the follicle the stream of cells destined to become the cuticle flatten circumferentially (Figure 1a) and tilt slightly so that they become interlocked with the cells of a follicle layer termed the inner root sheath. This interlocking serves to anchor the fiber to the skin of the animal.

When the shape change is nearing completion droplets of cuticle keratin are formed and these aggregate in the outer part of the cell, undergoing a transformation into a network (Figure 1b) which finally coalesces to yield the exocuticle layer (Figure 2) of the finished fiber (14). Concurrently with the formation of the exocuticle layer the cytoplasm and nucleus undergo a transformation and condense to form the endocuticle of the final fiber (Figure 2).

During fibrogenesis the plasma membranes surrounding the cells become extensively modified (10,11). Initially there is an increase in the number of desmosomes, which function as cell-to-cell attachment sites, and these are later replaced by a continuous layer of material between the cells. It is convenient to refer to the composite structure formed by this layer plus the

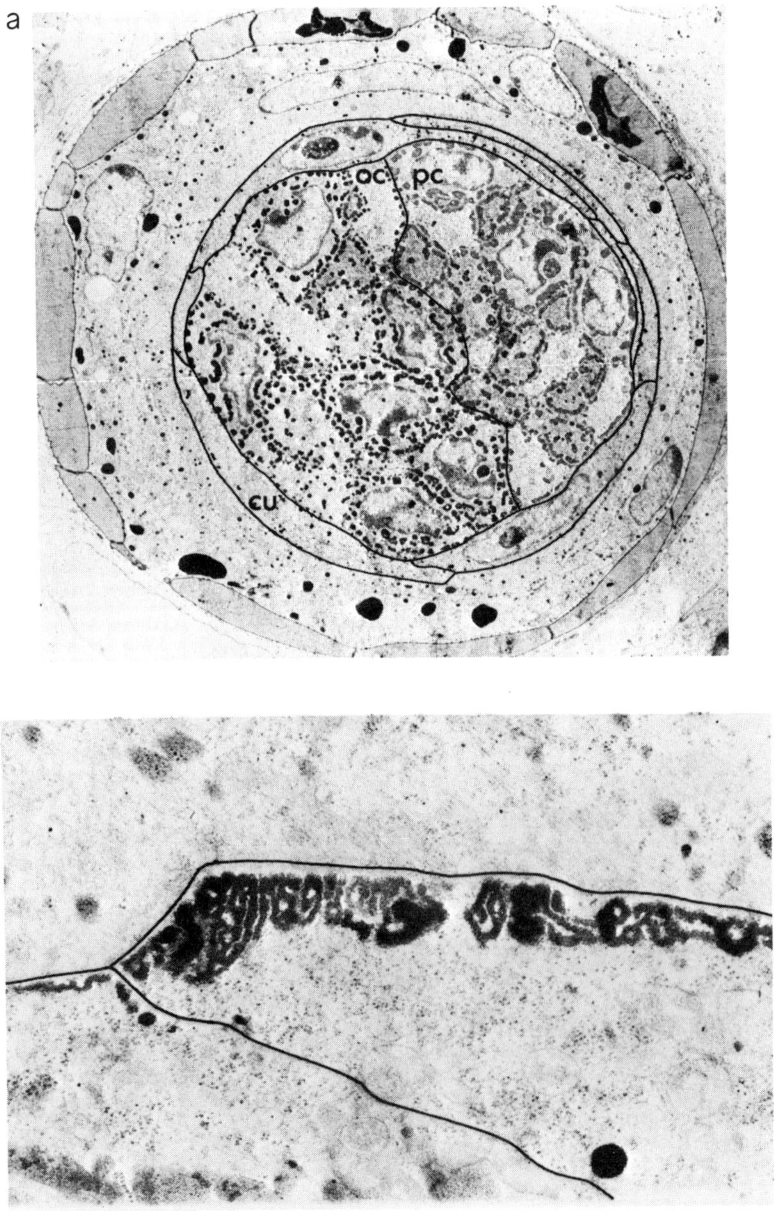

Figure 1. *a: Electron micrograph of cross-section through a wool follicle showing the developing wool fiber. Cells destined to become part of the cuticle (cu), paracortex (pc), and orthocortex (oc) are indicated.*
b: Network of lamellae that appear during formation of the exocuticle layer in cuticle cells. (*Reproduced, with permission, from Ref. 97. Copyright 1980, South African Wool & Textile Institute.*)

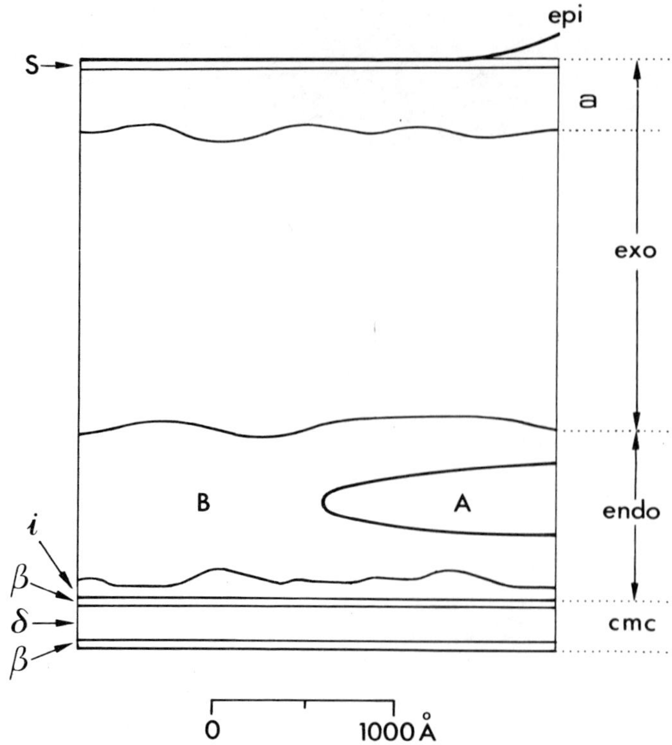

Figure 2. Diagram illustrating the structure of a mature cuticle cell. Key: epi, *epicuticle;* exo, *exocuticle;* endo, *endocuticle;* cmc, *cell membrane complex;* **β,δ**, *components of cell membrane complex;* **α**, *α-layer;* s, *ultra-cystine-rich layer;* A, *nuclear residue;* B, *cytoplasmic residue;* i, *cystine-rich inner layer. (Reproduced, with permission, from Ref. 97. Copyright 1980, South African Wool & Textile Institute.)*

modified plasma membranes as the cell membrane complex (Figure 2). There is evidence that the cell membrane complex between cuticle and cortical cells is different from that between apposed cuticle cells (11,15). The outer surface of the final fiber is enveloped in a very thin membrane termed the epicuticle (Figure 2) which is believed to be derived from the plasma membranes of the outer layer of cuticle cells.

The thickness of the cuticle varies widely being only a single cell thick in, for example, fine wools and up to 35 cells thick in, for example, pig bristle (9). The mechanical integrity of the cuticle appears to be enhanced by the occasional entrapment of a projection from one cell into a cavity in a neighbouring cell (16,17). The cells of the cuticle have a common structure, and the various layers that have been identified in mature cuticle cells are depicted diagrammatically in Figure 2. Pigment granules are not generally present in the cuticle cells of wool and human hair, but have been found in the cuticle cells of other types of hair (18).

The cuticle has a higher cystine content than whole wool (19) and cuticle-specific high-sulfur proteins have recently been isolated and characterized (20). In addition to disulfide cross linkages the proteins are thought to be further stabilized by the formation of isopeptide linkages between the sidechains of glutamyl and lysyl residues.

The bulk of the cystine, and presumably therefore the major part of the high sulfur proteins, is believed to be present in the exocuticle layer (21,22,23) and recently Woods and Orwin (14) have obtained evidence for the involvement of four types of keratin precursors in the formation of this layer. The "a"-layer of the exocuticle (Figure 2) appears to be richer in cystine than the remainder of the exocuticle and evidence has been obtained for the presence of a thin layer, situated at the outer surface of the exocuticle in the peripheral cells, of even higher cystine content (22).

The endocuticle (Figure 2) is derived by a process of consolidation, not presently understood, from the nucleus and cytoplasm of the developing cell. A technique based on susceptibility to enzymatic hydrolysis has been developed (23) to distinguish the endocuticle components derived from the nucleus (A) and the cytoplasm (B). The cystine content of the endocuticle is low (21,22,23) and mechanically the endocuticle is the weakest part of the cuticle (24). Although the endocuticle as a whole has a low cystine content there is a thin cystine-rich layer adjoining the cell membrane complex at the inner surface of the cell (8,25).

The cell membrane complex (Figure 2) is the end product of a series of little understood modifications of plasma membranes of apposed presumptive cuticle cells and of the material that is deposited between them during fibrogenesis (10,11). The central δ-layer (Figure 2) is often regarded as an "intercellular cement" and possibly contains some polysaccharide material (23,26).

Surrounding the δ-layer are two relatively inert β-layers which are presumed to contain an appreciable proportion of lipid, and to be derived from the plasma membranes of the developing cells. Attempts have been made to isolate and characterize the cell membrane complex of wool (27) and a protein component rich in glycyl, tyrosyl and glutamyl residues identified. This protein differs from the high-glycine-tyrosine proteins, discussed later, which can be extracted in large amounts from some types of hair and are believed to originate primarily from the cortex (3).

Electron histochemical studies suggest that there may be tyrosine-rich proteins present in the cuticle located intracellularly on either side of the cell membrane complex (9). The epicuticle (Figure 2) is only about 30Å in thickness (28,29) and accounts for only a very small proportion of the mass of a fiber (< 0.1%). Isolation and analysis of the epicuticle are therefore fraught with difficulty but both protein and lipid components have been identified and analyzed (29,30).

As mentioned earlier, the primary function of the ratchet-like profile of the cuticle is to anchor the fiber to the skin of the animal. As a result of this profile the coefficient of friction when the fiber is rubbed from root to tip is much less than when rubbed from tip to root (31,32,33). This directional friction effect is thought to assist in the expulsion of foreign matter from the pelt of the animal since relative motion between fibers will favour the motion of the foreign matter in a root to tip direction. In textile use the directional friction effect is exploited in the processes of fulling and felt making but also leads to the undersirable property of felting shrinkage in use. Shrink-resist treatments have been devised which either modify the profile or surface properties to reduce the directional friction effect, or form inter-fiber bridges which restrict inter-fiber movement (33,34).

The cuticle of hair, through its influence on such properties as friction and optical properties, exerts a profound influence on aesthetic appeal and such subjective properties as lustre, body, manageability, combability and style retention (32). Since the cuticle forms a continuous envelope around the fiber it also constitutes a barrier though which chemical reagents designed to react with the cortex must pass.

The Cortex. In most hairs the cortex constitutes the main bulk of the fiber and so plays a major role in determining bulk mechanical properties and chemical reactivity. During their movement through the follicle the cells destined to form the cortex elongate parallel to the fiber axis. In the initial stages the number of desmosomes increases as does the number of gap junctions, which are thought to be involved in the transport of metabolites between cells (35). Concurrently bundles of filaments termed macrofibrils form at the cell boundaries, frequently in association with the desmosomes (12,36).

As the macrofibrils increase in length they become preferentially oriented parallel to the direction of growth and two distinct behaviours are found in the latter stages of development (Figure 3). In some cells (para cells) the macrofibrils tend to fuse, whilst in others (ortho cells) the macrofibrils remain discrete (16,36,37). In many fine fibers the two types of cell are grouped to form distinct strands (Figure 1a) which appear to be associated with crimp formation (8). Coarser fibers exhibit less definite arrangements of cells and the ortho and para characteristics are less well developed. Cells of intermediate character are also present in some fibers (38,39).

Remnants of the nuclei of developing cortical cells can clearly be recognized in the final fiber but the cytoplasmic debris is distributed throughout the cell. In ortho cells the cytoplasmic debris forms an intermacrofibrillar network, whilst in para cells the deposits are more discrete (16,37). The cell membrane complex between cortical cells exhibits a layer structure similiar to that shown in Figure 1a for cuticle cells but several lines of evidence indicate that they differ in several important respects (15,16,35,37,40). However the structural bases of these differences are not presently understood.

The macrofibrils in the cortex are entirely intracellular and so the cell membrane complex provides both a mechanical linkage between the keratin deposits in adjacent cells and a potential barrier to the entry of chemical reagents into the keratin deposits. The importance of the cell membrane complex to the mechanical continuity of hair was demonstrated by Elöd and Zahn (41), who treated hair with a proteolytic enzyme which preferentially digested the cell membrane complex. The tensile strength decreased rapidly and after a few days the fibers disintegrated, with the release of cortical cells. The mechanical coherence of the cortex is also enhanced by the interdigitation of cortical cells during fibrogenesis (16). Processes extending from the peripheral cortical cells also interlock with the cuticle (17).

The Medulla. Hairs commonly, though not universally, have a medulla which is derived from a central stream of cells that become vacuolated during fibrogenesis. These cells produce granules which transform into a coating around the periphery of the vacuoles. A protein believed to derive from the granules has been isolated and purified and found to be rich in arginine and glutamic acid plus glutamine (42). During maturation of the medulla ε-(γ-glutamyl) lysine cross links are formed (43). These isopeptide linkages are responsible for the extreme insolubility of the mature medulla, which is almost devoid of the cystine cross linkages responsible for the insolubility of the keratin deposits in the cortex.

The large change of refractive index at the boundary between the main body of the fiber and the vacuoles of the medulla give medulated fibers special optical properties which are utilized in

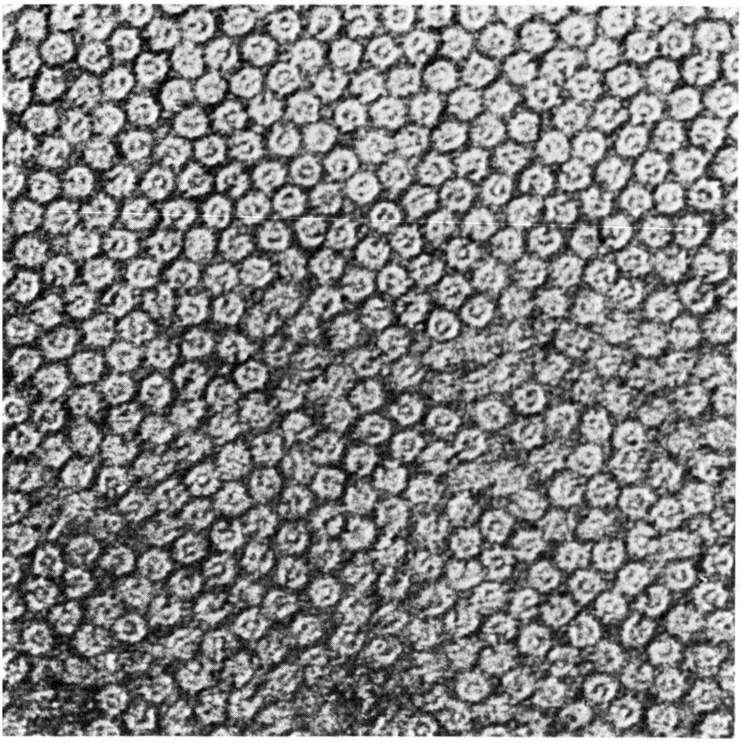

Figure 3. Electron micrograph of a cross-section of paracortex of a Merino wool fiber showing the microfibril–matrix complex.

producing specific fabric properties. Since scattering is greatest for the blue end of the spectrum this offsets the natural yellow-brown colour of unpigmented fibers, making them appear whiter. Medulated fibers play an important part in pelt structure since the medulla effects an economy of material and weight, and improved thermal insulation properties without significant loss of bending strength (5).

Molecular Texture and Composition

Cuticle Cell Keratin. Studies of the composition of cuticle cells are complicated because of the difficulties encountered firstly in isolating undamaged cells and secondly in solubilizing the contents. The most recent studies (20) yielded an extract, representing about 30% of the mass of the cuticle cell preparation which contained keratin proteins in which for every 100 residues there were on average around 18 cysteinyl, 15 seryl, 12 prolyl and 8 glutamyl residues. These high-sulfur proteins have a composition which differs from that of the high-sulfur proteins of the cortex, discussed later (Table I), and appear to be cuticle-specific. Woods and Orwin (14) obtained evidence for four distinct types of keratin deposit, one fibrous and three globular but no correlation between these deposits and the extracted proteins is possible on the basis of present knowledge.

When examined by X-ray or electron diffraction cuticle cells give patterns characteristic of polypeptide chains devoid of regular secondary structure (44,45). This observation is in accord with the high content of prolyl residues which, unless precisely spaced along the polypeptide chain, do not favour the formation of regular conformations. The early view that the exocuticle is formed simply by the coalescence of amorphous droplets of high-sulfur proteins must however be revised in the light of recent studies by Woods and Orwin (14) which show that an ordered lamellar network around 600Å in thickness is formed during the development of the exocuticle (Figure 1b).

Composition of Cortex Keratin. The keratin deposits in the cortex (macrofibrils) have a composite structure (Figure 3) in which microfibrils around 75Å in diameter are embedded in a cystine-rich matrix (12,16,22,37). The microfibrils appear to be related to the so-called intermediate filaments which form part of the intracellular cytoskeleton in a number of diverse tissues (46). Intermediate filaments generally have the form of a hollow tube around 70-100Å in diameter but the microfibrils in cortical cell keratin have a central core which may be a specific adaption peculiar to keratin.

Three distinct groups of proteins have been extracted from keratin fibers (Figure 4): a low-sulfur group, a high-sulfur group and a group rich in glycyl and aromatic residues (high-tyrosine, high-glycine-tyrosine, or HGT group). This latter group

Table I
Amino acid compositions of various protein fractions isolated from Merino wool (Residues per 100 residues).

Residue	Low-sulfur fraction[a]	High-sulfur fraction[a]	HGT Type I fraction[a]	HGT Type II fraction[a]	Cuticle extract[b]
Alanine	6.9	2.9	1.5	1.1	5.9
Arginine	7.3	5.9	5.4	4.7	5.0
Aspartic acid	9.0	3.0	3.3	1.8	2.5
Cysteine	6.0	18.9	6.0	9.8	18.1
Glutamic acid	15.7	8.4	0.6	0.7	8.1
Glycine	7.7	6.9	27.6	33.6	6.4
Histidine	0.6	0.8	1.1	0.1	0.5
Isoleucine	3.6	3.6	0.2	0.2	1.8
Leucine	10.2	3.9	5.5	5.3	5.8
Lysine	3.5	0.6	0.4	0.4	1.3
Methionine	0.6	0.0	0.0	0.0	0.1
Phenylalanine	2.5	1.9	10.3	4.5	1.4
Proline	3.8	12.5	5.3	3.0	12.4
Serine	8.2	12.7	11.8	10.9	15.1
Threonine	4.8	10.3	3.3	1.7	5.1
Tyrosine	3.6	2.1	15.0	20.3	3.0
Valine	6.1	5.6	2.1	1.4	7.2

[a]Taken from Gillespie (3)

[b]Taken from Ley and Crewther (20). Traces of tryphophan, lanthionine and allo-qsoleucine residues were also detected.

[c]Includes asparagine residues.

[d]Includes glutamine residues.

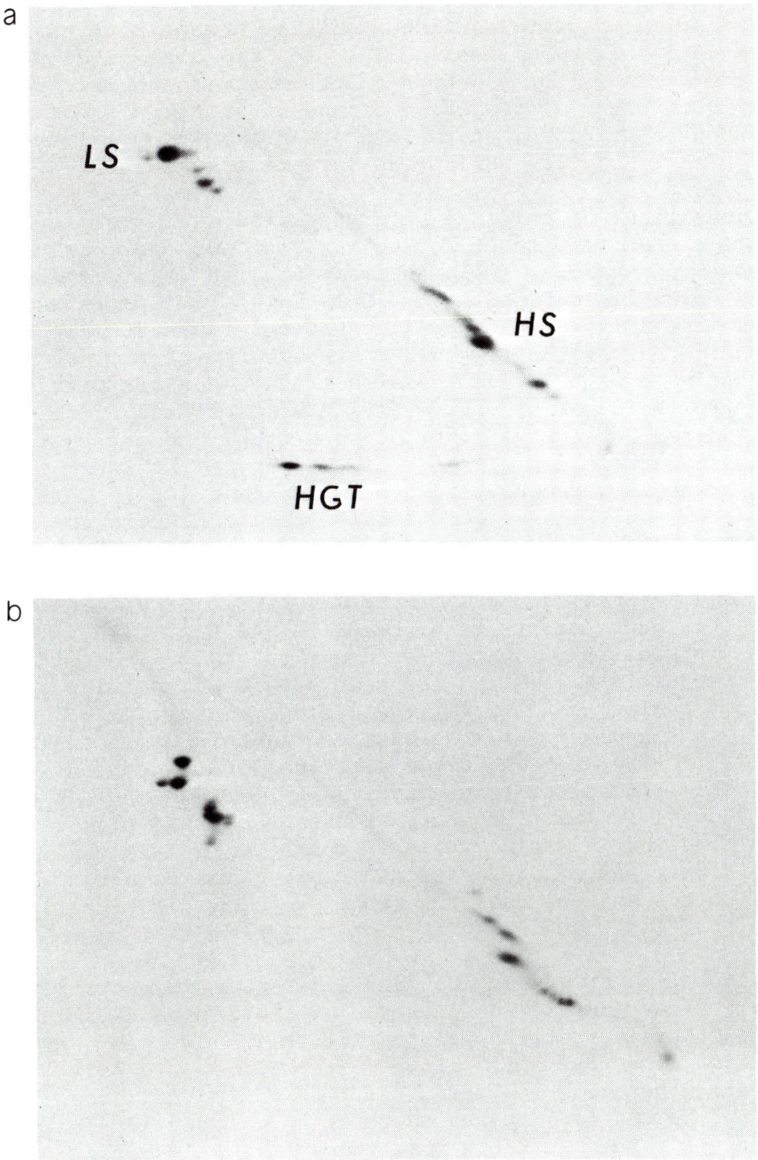

Figure 4. Two-dimensional polyacrylamide gel electrophoretic patterns of soluble proteins extracted from mouse hair (a) and human hair (b). Three groups of proteins are present in a: low-sulfur group (LS), high-sulfur group (HS), and high-glycine-tyrosine group (HGT). In b the HGT group is virtually absent.

contains two distinct fractions which have been designated Type I and Type II (Table I). The isolation, purification and characterization of proteins from the three groups have been described in detail by Gillespie (3). An important finding has been that keratin is not a substance of constant composition: the relative proportions of the three groups of proteins are influenced by nutritional status and other physiological factors. Major species to species variations have also been found (Figure 5).

Various lines of evidence have suggested that the microfibrils are composed of low-sulfur proteins and this has been demonstrated directly by isolating microfibrils from developing cells and solubilizing them (47). Only low-sulfur proteins were detected. The amino acid composition of the isolated microfibrils also matched that of the low-sulfur protein fraction.

An estimate of the mean cross-sectional area per microfibril in the macrofibril can be obtained from measurements of the low-angle X-ray equatorial diffraction pattern and early studies showed that the area per microfibril, and hence the intermicrofibrillar distance, increased with cystine content (48). This suggested that the high-sulfur proteins were concentrated in the matrix. Later it was shown that the volume fraction of matrix in dry keratin, calculated from estimates of the microfibrillar diameter and the intermicrofibrillar distance, correlated well with the combined contents of high-sulfur and high-glycine-tyrosine proteins (49). Direct evidence for the presence of high-sulfur proteins in the matrix has been obtained from electron microscope studies using specific staining methods (22).

All wools and hairs so far examined contain appreciable quantities (> 15%) of high-sulfur proteins but the content of high-glycine-tyrosine proteins may be very low (50) as for example in Lincoln wool (< 1%). In contrast the content of high-glycine-tyrosine proteins in Merino wools is appreciable (about 13%) and amino acid analyses of preparations of ortho and of para cells from the cortex suggest that the ortho cells have a higher content of these proteins than do the para cells (51).

Conformation of Cortex Proteins. Keratin fibers yield high-angle X-ray diffraction patterns of the α-type (8,52) indicating that in part the polypeptide chains have an α-helical conformation, and that small groups of α-helices are twisted around each other to form what are termed coiled-coil ropes (53). Birbeck and Mercer (12) suggested that the polypeptide chains responsible for the α-pattern were those of the low-sulfur protein and were located in the microfibrils. This view is supported by the observation that the S-carboxymethyl derivatives of extracted low-sulfur proteins have a partly α-helical conformation in solution whereas the corresponding derivatives of the extracted high-sulfur and high-glycine-tyrosine proteins are virtually devoid of regular secondary structure in solution

(51,54,55). Some caution is necessary in extrapolating from solution studies of the S-carboxymethyl derivatives of the latter two types of protein to the situation in the native fiber since the cysteine residues, which are present in appreciable proportions, have been reduced and modified by the addition of a negatively charged group to the sidechain.

Fragments rich in α-helix have been isolated from unfractionated preparations of the S-carboxymethyl derivatives of the low-sulfur proteins after partial enzymatic hydrolysis (55). Oriented films cast from solutions of the fragments give an α-type X-ray pattern indicating that the coiled-coil conformation is present, and several orders of a 160Å meridional pattern are present and this has been correlated with the end-to-end distance of the fragments (56).

Two types of chain segment have been isolated from the helix-rich fragments: the first, designated Type I, contains 103 residues and the second, designated Type II, contains 109 residues (55). Crick (53) pointed out that the sidechains of the residues in the interface between α-helices in a coiled-coil rope would be expected to have a hydrophobic character since they are shielded from contact with the aqueous solvent. The geometrical form of the coiled-coil is such that there exists a repeating pattern of seven residues, and if consecutive residues in the heptapeptide pattern are designated a - g, as in Figure 6, the sidechains of residues a and d would be predicted to be predominantly non-polar in character. No specific prediction can be made for the remaining residues. The amino-acid sequences of the Type I and Type II segments have been determined (57,58) and both exhibit a pattern of the predicted type.

Crick (53) described two coiled-coil rope models: a two-strand rope which has been shown to be present in a number of muscle proteins (52), and a three-strand rope which he suggested was present in wool and hair. On the basis of their studies of helix-rich fragments isolated from wool, Crewther and Dowling (59) suggested that each fragment consisted of a section of triple helix some 128Å in length with a small amount of random chain at each end. Early studies of the X-ray diffraction pattern also favoured the three-strand rope model (60) but later it became clear that direct analysis of the diffraction pattern in terms of that predicted for a coiled-coil rope was complicated by superposed interference functions from higher levels of structure (61). McLachlan (62) studied the sequences of the Type I and Type II segments in relation to the way in which they could pack in the two coiled-coil models and concluded that the two-strand rope model is the more likely for wool and hair. Possible interactions between Type I and Type II segments in a two-strand rope (63) and in a three-strand rope (64) have been studied but neither was found to be significantly favoured. A clear choice between the two-strand and three-strand rope models does not seem to be possible at present on the basis of the physicochemical or

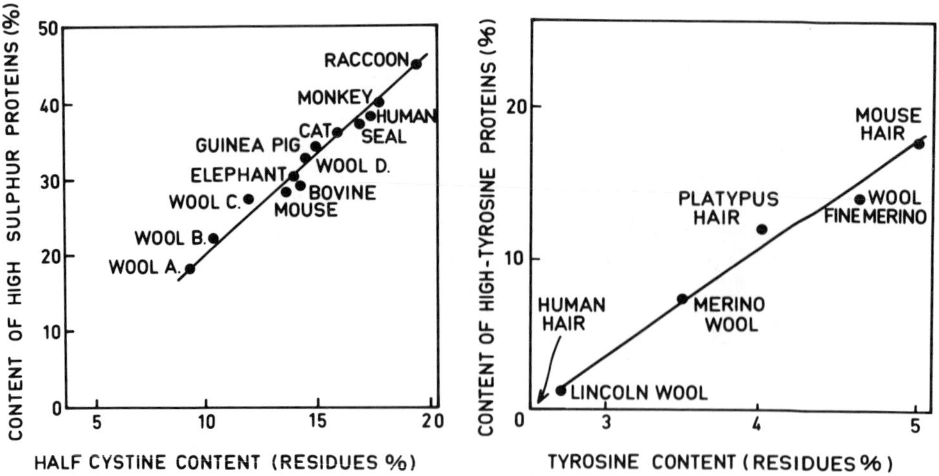

Figure 5. The contents of high-sulfur proteins (left) and high-glycine-tyrosine proteins (right) in keratin fibers, illustrating variability. Reproduced, with permission, from Ref. 4. Copyright 1974, Pergamon Press.

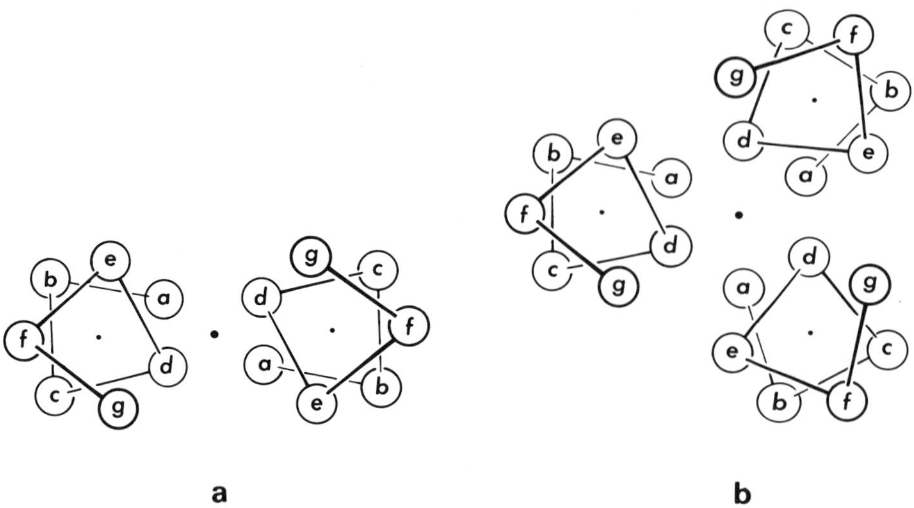

Figure 6. Helical projections illustrating the interactions between residues in the two-strand coiled-coil rope (a) and the three-strand coiled-coil rope (b). In both cases the residues designated by a and d are located in the interior of the rope. (Reproduced, with permission, from Ref. 97. Copyright 1980, South African Wool & Textile Institute.)

diffraction data currently available. Very little is known about the conformation of the non-α-helical portions of the low-sulfur proteins. The frequency with which prolyl and glycyl residues occur (65) suggest however that the polypeptide chain is highly convoluted.

Films prepared from the S-carboxymethyl derivatives of the high-sulfur proteins exhibit evidence of the presence of a β-conformation after certain preparative treatments (66), but as discussed earlier, caution is needed in extrapolating these findings to the situation in the native fiber. Certain of the high-sulfur proteins have been found to contain repeating patterns of residues (67-71) and there is evidence for both pentapeptide and decapeptide patterns (54,72,73). The pentapeptide has a favoured composition

```
                Gln          Ser
- Cys - Cys - Arg - Pro - Thr -
   1     2     3     4     5
```

and the repetition of this pattern implies a repetition of conformation. If completely regular this repetition would lead to a helical structure, but the available evidence suggests (74) that it is more likely that it is the local conformation that is common to each pentapeptide. The sequence

```
- Gln/Arg - Pro - Ser/Thr - Cys -
     3       4       5      1(+5)
```

is predicted to have a high probability of existing in a β-bend conformation (75) and there is also a considerable body of indirect evidence which suggests that an appreciable proportion of the disulfide cross linkages are formed intramolecularly (44,54,55). Taken together these two facts suggest that the preferred pentapeptide sequence has a local conformation that consists of a β-bend stabilized by a disulfide cross link as follows:

```
4 Pro - Y 5
|       |
3  X...Cys 1(+5)
|     /  |
2 Cys
|
```

where the dotted line represents a hydrogen bond and the broken line represents a disulfide linkage. A conformation of this type would lead to sequence-dependent properties since the presence of Cys residues in both positions 2 and 1(+5) would give a highly stable intramolecularly-bonded loop whilst the absence of Cys from one of these positions would leave an unpaired thiol group in either the same or a different molecule.

High-sulfur proteins that exhibit the pentapeptide repeat also have chain seqments in which the repeat is absent and the composition is distinctly different (67). In the SCMKB-2 series of high-sulfur proteins for example there is a central segment of 27 residues which is devoid of Cys residues and rich in Gly residues which it has been suggested (67) forms a flexible link between the two segments which demonstrate the pentapeptide repeat. High-sulfur proteins of the BIIIB family (76) do not exhibit any obvious evidence of sequence regularity and a possible series of intrachain disulfide linkages involving multiple folding has been suggested by Lindley (54).

The sequences of two Type I high-glycine-tyrosine proteins have been reported (77,78), the first containing 62 residues and the second 84. Little obvious homology exists between the sequences but in both cases there is a strong tendency for glycyl residues to be followed by hydroxyl-containing residues (seryl, threonyl or tyrosyl residues). In the smaller protein there is a repeating pattern of the type (-Gly-X-) which is a feature of the sequence in the crystalline regions of certain arthropod silks (52). This suggests that a part of the molecule may have a β-conformation similar to these silks.

Spatial Arrangement of Cortex Proteins

As mentioned earlier there is good evidence that the microfibrils are composed of low-sulfur proteins. In electron micrographs of cross sections of wool and hair that have been reduced and treated with heavy-atom stains the microfibrils are revealed as circular areas which have taken up less stain than the surrounding matrix (16). The diameters of the less densely stained areas are difficult to assess but appear to be in the range 60-80Å depending upon the particular staining procedures that have been used. The diameters of isolated microfibrils are also in this range (47,79,80) and it seems likely therefore that the circular areas seen in cross-section account for the major part of the low-sulfur protein. It must be remembered however that the low-sulfur proteins have chain segments of composition similar to that of the high-sulfur proteins and if any of these segments projected outside the circular area they would merge with the intensely stained background. Similarly small projections with an α-helical structure would not be preserved or resolved using current procedures. Several lines of evidence suggest the possibility that part of the low-sulfur proteins may project beyond the circular area which is the current operational definition of the microfibril in electron microscope studies (44,81).

Some substructure is visible within the circular area and initially this was thought to resemble the 9 + 2 arrangement of filaments found in cilia and flagella (82). The term protofibril was used to describe these supposed sub-filaments and a number of authors claimed to have demonstrated isolated protofibrils in the electron microscope. Later however it was established that the protofibrils were in fact derived from cellulosic contaminants (83). When the substructure revealed by electron microscopy is recorded close to focus a "ring-core" appearance is obtained (84). The lightly-stained ring has a radius of about 30Å and X-ray diffraction studies suggest that it is composed principally of segments of coiled-coil rope (85). The lightly-stained core may also be composed of coiled-coil ropes but there is no definitive evidence on this point and it could equally-well be composed of hydrophobic non-helical parts of the low-sulfur protein molecules that are inaccessible to the electron stain. It may be recalled that it is the core portion of the microfibril that appears to differentiate it from other intermediate filaments. The annulus between the ring and core takes up heavy atom stain and is thought to correspond to non-helical portions of the molecule, which are known to be sulfur-rich (65, 86).

Dimensions determined from electron microscope studies are prone to error because of dimensional changes which result from the fixation and staining procedures normally employed but an alternative method of estimating microfibril diameter has been devised, based on the assumption that in the native fiber there will be a discontinuity in electron density between the microfibrils and the matrix (87). In addition to the high-angle X-ray pattern of the α-type, indicative of the presence of coiled-coil ropes, keratin fibers exhibit a low-angle pattern that arises from regularities in the spatial arrangement of the keratin proteins of the cortex. The equatorial low-angle pattern contains information about the projection of the structure down the fiber axis and the intensities and profiles of the three low-angle intensity maxima that are present are determined both by the distribution of electron density in the microfibril and by the nature of the microfibril packing within the macrofibrils. If the outer maxima and minima are fitted to the scattering function predicted for a cylinder, values of cylinder diameters in the range 75 ± 3Å are obtained for various dried keratinous tissues (49).

The microfibrils in keratin fibers are preferentially oriented parallel to the length of the fiber but the lateral packing arrangement exhibits a variable degree of ordering and differs in the ortho and para types of cortical cell (16). In the para-type cells the lateral packing has a hexagonal character whilst in ortho-type cells the macrofibrils consist of concentric layers of microfibrils in which the inclination of the microfibrils to the fiber axis increases progressively with distance from the centre of the macrofibril. In electron micrographs of stained wool fibers which contain both ortho and para-type cells it has been noted that the microfibrils in the para cells are separated by greater distances than are the microfibrils in the ortho cells (88,89) and a similar conclusion was reached from electron diffraction studies of stained cross sections of wool (90,91). In both cases the results were interpreted as indicating that the para cells contained a higher proportion of matrix than the ortho cells but these conclusions must be treated with caution since the separation between microfibrils is known, from X-ray studies, to be increased by stain deposition (48), and the reactivities to stain may be different in the two types of cell.

The meridional X-ray diffraction pattern of wool and hair contains information about the distribution of electron density in the direction of the fiber axis. The pattern consists of discrete reflections indicating that the distribution is periodic and early studies suggested that the period was 198Å (92,93), but more recently it has been found that a period of 470Å is required to index the presently available data (94). The 5.16Å reflection, which is characteristic of the coiled-coil conformation, indexes as an order of 470Å and so it follows that the distribution of low-sulfur proteins, and hence the microfibril, is axially periodic.

The off-meridional reflections have been catalogued in detail (85) and shown to be consistent with the presence of a helical structure in which the basic helix has a unit height of 470Å and a unit twist of around 49°. There is some evidence (85) that within each structural unit there is a pseudo-repeat with a unit height of 470/7 Å and a unit twist of around 109° (Figure 7). Associated with each pseudo-repeat of structure there will be a group of low-sulfur proteins which occupy a volume of approximately

$$V = 2.93 \times 10^5 \text{ Å}^3$$

in the dry fiber (85) and if the density is assumed to be 1.3 g cm^{-3}, the associated molecular mass will be

$$M = 2.29 \times 10^5 \text{ daltons.}$$

Without further evidence no restriction can be placed on the spatial form of the volume excluded by this group of low-sulfur proteins. In the illustration in Figure 7 there are three strings of pseudo-repeats 201Å in length. An alternative arrangement suggested by Ahmadi *et al.* (95) consists of ten strings of pseudo repeats 650Å in length. In both instances the volume per pseudo-repeat will be the same and equal to V.

Two types of protein chain are present in the low-sulfur group of proteins: these are termed the Component 5/Component 7 (C5/7) and Component 8 (C8) groups. In Merino wool, the only material that has been studied in detail so far, the two types of component appear to be present in approximately equal proportions (95) and to have chain weights of approximately 58,000 and 50,000

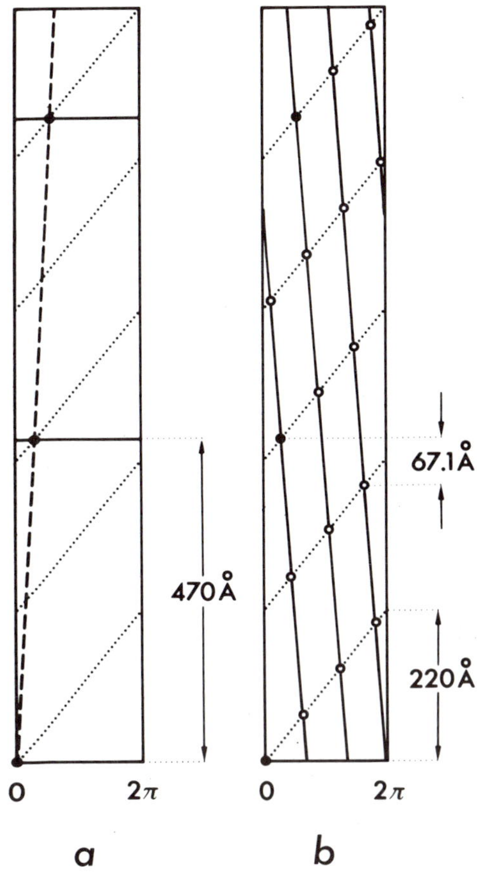

Figure 7. Radial projections of the microfibril helix (a) (●,---) and a pseudo-repeat suggested by the X-ray diffraction data (b) (O ....). Three strings of pseudo-repeats are indicated (—).

respectively (96). Thus the mean chain weight would be 54,000 and the estimated number of chains in the pseudo-repeat would be

$$n = M/54,000 = 4.24$$

This suggests that the pseudo-repeat has the composition $(C5/7)_2$ $(C8)_2$ proposed by Ahmadi *et al*. from chemical studies (95). If the suggestions, mentioned earlier, that the low-sulfur proteins project outside the cylindrical volume delineated by stain in electron micrographs are true then the value of V calculated earlier will be too low and the composition $(C5/7)_3$ $(C8)_3$ ought also to be considered (97). The microfibril diameter predicted for the $(C5/7)_2$ $(C8)_2$ model is 72.4Å which is close to the value of 74.5Å estimated from X-ray measurements (49); the diameter predicted for the $(C5/7)_3$ $(C8)_3$ model is 88.6Å which is equivalent to a 7Å coating of readily stainable material around each microfibril.

On the basis of the data available at present the chain composition $(C5/7)_2$ $(C8)_2$ for the pseudo-repeat of structure appears to be the most plausible choice. A disquieting fact however is that Steinert *et al*. (98) have obtained strong evidence for a structural unit in the microfibrils of the so-called soft keratins which contains three, or some multiple of three, chains. In view of the many similarities between the soft keratins and the hair keratins such as wool and hair, this could be argued as favouring *a priori* a pseudo-repeat in wool and hair that contained three or some multiple of three chains. Clearly further chemical and physicochemical studies of the subunit structures of wool and hair are needed.

Very little is known about the arrangement of the proteins in the matrix. The positions of certain of the off-meridional maxima in the X-ray diffraction pattern indicate that the helical symmetry present in the microfibril extends to include parts of the structure outside the lightly stained circular areas seen in electron micrographs (94). This suggests either that projections of the low-sulfur proteins exist in densely stained areas, or that the matrix proteins are, in part, ordered as regards their disposition with respect to the microfibril. A number of staining procedures which would be expected to react preferentially with the matrix proteins have been shown to affect the intensities of the low-angle X-ray reflections (52,99,100); this also supports the idea of a matrix which is at least in part ordered (99,100, 101).

The Influence of Molecular Texture on Properties

The bulk of the keratin in wool and hair has a microfibril-matrix texture and thus belongs to the class of materials termed composites (102). The microfibrils are preferentially aligned parallel to the fiber axis and so the fiber exhibits anisotropic behaviour in many of its physical properties.

Sorption and Swelling. The dimensional changes that occur when a fiber such as hair is taken from the dry to the wet state are markedly anisotropic (103); the increase in diameter is around 14%, whereas the increase in length is less than 2%. X-ray diffraction studies reveal a parallel behaviour of the micro-fibril-matrix complex in that the axial repeat of structure changes only slightly (92), whereas the intermicrofibrillar distance increases by about the same proportion as the diametric swelling (8). Two conflicting views exist about the distribution of the sorbed water within the fiber. Measurements of the spacings of intensity maxima in the X-ray diffraction pattern associated with the internal structure of the microfibril and with the intermicrofibrillar distance (8) suggest that the bulk of the sorbed water is accommodated in the matrix. The volume swelling of the matrix was estimated to be around 50%, whilst the volume swelling of the microfibril was estimated to be only about 10%. The changes that occur in mechanical properties when a dry fiber is wetted also support the notion that the matrix swells much more than the microfibril (6,104).

An alternative view, put forward by Spei and Zahn (105), is based on the observation that there is a negative correlation between fiber swelling and matrix content (105,106). Since fibers with a high proportion of microfibrils swell more than fibers with a low proportion of microfibrils it was argued that the micro-fibrils must be responsible for most of the swelling. However an alternative explanation of the negative correlation between fiber swelling and matrix content has been suggested, based on the mode of assembly of the microfibril-matrix complex during fibrogenesis (5). A curious feature of the assembly process is that similar packing arrangements of microfibrils are found in materials in which both the amount and nature of the intervening matrix vary over wide limits (107). This prompted the suggestion (5) that there is an initial step in the assembly process in which the lattice is formed and stabilized either by portions of the low-sulfur proteins that project from the microfibrils or by small amounts of "essential" high-sulfur proteins that bind specifically to the newly-assembled microfibrils prior to their incorporation in the lattice (Figure 8). The intermicrofibrillar distance in the wet state thereby becomes fixed and the subsequent incorporation of the matrix in the spaces between the microfibrils displaces water. Clearly the amount of shrinkage on drying (and therefore swelling on rewetting) would be least in fibers of greatest matrix content. An observation which supports this concept (108) is that wool fibers treated with ninhydrin exhibit a negative correlation between amount of ninhydrin taken up by the fiber and the amount of water sorbed by the treated fiber.

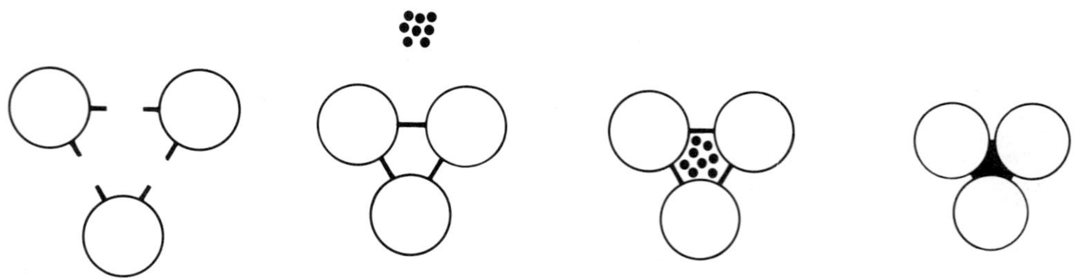

Figure 8. Diagram illustrating a possible mechanism of assembly of the microfibril–matrix complex; Key: a and b, initial stage in which microfibrils assemble at center-to-center distances determined by projections; c, entry of matrix protein molecules; and d, shrinkage upon drying, which will be negatively correlated with the amount of matrix protein deposited. (Reproduced, with permission, from Ref. 97. Copyright 1980, South African Wool & Textile Institute.)

Mechanical Properties. The microfibril-matrix texture has a profound effect on the mechanical properties of keratin fibers (5,6). In the dry state the mechanical properties of the two components are very similar but in the wet state the swollen matrix is extremely weak (104) and this results in anisotropic properties in the hydrated state (6). It is difficult to measure transverse tensile moduli in fibers but in porcupine quill, which has a related structure, the transverse tensile modulus is only 0.3 GPa (109) compared with a value of 3.5 GPa (110) for the longitudinal tensile modulus.

As discussed earlier, both the composition and the amount of matrix show wide variations in different fibers and a correlation has been established between the proportion of matrix in keratinous materials and the transverse compression modulus in the wet state (111,112). This correlation supports the notion of intermicrofibrillar linkages, as discussed earlier, since exclusion of water by the addition of matrix proteins would be expected to give a stiffer matrix (112,113).

Composites in which linearly-elastic, high-modulus filaments are embedded in a viscoelastic matrix of lower modulus have important properties not possessed by the individual components. In particular the matrix tends to distribute applied stress evenly over the filaments thus preventing the propagation of cracks from local imperfections or damaged areas. A side-effect of the fact that the microfibril-matrix complex is assembled and cross-linked in the hydrated state is that when subsequently exposed to an unsaturated environment the matrix contracts and so the microfibrils are subject to a small amount of longitudinal compression. Thus in an unsaturated environment the fiber is "pre-stressed" (6).

Chemical Properties. The important part played by the molecular texture of a fiber in determining its chemical propies has been discussed by Shutz (114). In common with many other natural composites the highly-ordered regions of a keratin fiber (microfibrils) are relatively physicochemically inert and the chemical properties of the fiber are largely determined by the less-well ordered regions (matrix) and their interface with the highly-ordered regions. In the case of hydrated keratin fibers the bulk of the ring portion of the microfibril and the material enclosed by it would be expected to be relatively inaccessible, whereas the highly swollen matrix and the surfaces of the microfibrils would be expected to be readily accessible to reagent molecules of appreciable dimensions. Support for this view is obtained from the fact that a great variety of chemical reagents, including dyes, can be introduced into keratin fibers without disturbing the perfection of the parts of the X-ray diffraction pattern thought to be associated with the framework of the microfibril.

The helical segments of the low-sulfur proteins are

particularly rich in charged residues and it seems likely that an appreciable fraction of these sidechains will be located on the outer surface of the ring portion of the microfibril and thus be accessible to reagents which can penetrate the matrix. There is evidence that the microfibril-matrix texture places a physical limitation on the size of molecules than can move freely through the matrix (32,115). This limiting size is likely to be increased in the presence of reagents such as formic acid, which partially disrupt the native structure (113).

If the linked-microfibril scheme depicted in Figure 8 is, in fact, present it would provide an explanation for the observation that the relative rate of diffusion of the dye Orange II into different fibers is negatively correlated with sulfur content (116). The content of high-sulfur proteins is proportional to the sulfur content (3) and so the reduced rate of diffusion is readily understood in terms of the increased occupancy of the space between the microfibrils.

Acknowledgements

We wish to thank Drs. P.A. Tulloch and E.F. Woods for their helpful comments on the first draft of this contribution. Also, we thank G.E. Rogers and R.C. Marshall for use of figures.

Literature Cited

1. Harris, M. Textile Bulletin 1938, 54, 6-8.
2. Hunter, L. in "Proc. 6th Int. Wool Text. Res. Conf. Pretoria"; Vol. 1, pp. 133-193, South African Wool and Textile Research Institute: Pretoria, 1980.
3. Gillespie, J.M. in "Biochemistry and Physiology of Skin"; (Goldsmith, L.A. ed.) Oxford University Press: Oxford, 1981, in press.
4. Gillespie, J.M.; Frenkel, M.J., Comp. Biochem. Physiol. 1974, 47B, 339-346.
5. Fraser, R.D.B.; MacRae, T.P. in "The Mechanical Properties of Biological Materials"; (Vincent, J.F.V. and Currey, J.D. eds.) S.E.C. Symposium No. 34, pp. 211-246, Society for Experimental Biology: London, 1980.
6. Bendit, E.G.; Feughelman, M. in "Encyclopedia of Polymer Science and Technology"; Vol. 8, pp. 1-44, Wiley: New York, 1968.
7. Chapman, B.M. J. Text. Inst. 1969, 60, 181-207.
8. Fraser, R.D.B.; MacRae, T.P.; Rogers, G.E. "Keratins, Their Composition, Structure and Biosynthesis"; Charles C. Thomas: Springfield, Ill, 1972.
9. Swift, J.A. in "Chemistry of Natural Protein Fibers"; (Asquith, R.S. ed.) pp. 81-146, Plenum Press: New York, 1977.
10. Orwin, D.F.G. Int. Rev. Cytol. 1979, 60, 331-374.
11. Orwin, D.F.G. in "Fibrous Proteins: Scientific, Industrial and Medical Aspects"; (Parry, D.A.D. and Creamer, L.K. eds.) Vol. 1, pp. 271-297, Academic Press: London, 1979.

12. Birbeck, M.S.C.; Mercer, E.H. J. Biochem. Biophys. Cytol. 1956, 3, 203-214.
13. Swift, J.A. Nature (London) 1964, 203, 976-977.
14. Woods, J.L.; Orwin, D.F.G. in "Fibrous Proteins: Scientific, Industrial and Medical Aspects"; (Parry, D.A.D. and Creamer, L.K. eds.) Vol. 2, pp. 141-149, Academic Press: London, 1980.
15. Nakamura, Y.; Kanoh, T.; Kondo, T.; Inagaki, H. in "Proc. 5th Int. Wool Text. Res. Conf. Aachen, 1975"; Vol. II, pp. 23-32, Deutsches Wollforschungsinstitut: Aachen, 1976.
16. Rogers, G.E. J. Ultrastruct. Res. 1959, 2, 309-330.
17. Piper, L.P.S. J. Textile Inst. 1966, 57, T185-T190.
18. Piper, L.P.S. Nature (London) 1967, 213, 596.
19. Bradbury, J.H.; Chapman, G.V.; King, N.L.R. in "Symposium on Fibrous Proteins, Australia, 1967"; (Crewther, W.G. ed.) pp. 368-372, Butterworths: Australia, 1968.
20. Ley, K.F.; Crewther, W.G. in "Proc. 6th Int. Wool Text. Res. Conf. Pretoria"; Vol. II, pp. 13-28, South African Wool and Textile Research Institute: Pretoria, 1980.
21. Bradbury, J.H.; Ley, K.F. Australian J. Biol. Sci. 1972, 25, 1235-1247.
22. Dobb, M.G.; Murray, R.; Sikorski, J. J. Micr. 1972, 96, 285-299.
23. Swift, J.A. in "Proc. 5th Int. Wool Text. Res. Conf. Aachen, 1975"; Vol. II, pp. 12-21, Deutches Wollforschungsinstitut: Aachen, 1976.
24. Orwin, D.F.G.; Thomson, R.W. in "Proc. 5th Int. Wool Text. Res. Conf. Aachen, 1975"; Vol. II, pp. 173-182, Deutches Wollforschungsinstitut: Aachen, 1976.
25. Swift, J.A. J. Roy. Micr. Soc. 1967, 88, 449-460.
26. Orwin, D.F.G. Australian J. Biol. Sci. 1970, 23, 623-635.
27. Bradbury, J.H.; Chapman, G.V.; King, N.L.R. "Proc. 3rd Int. Wool Text. Res. Conf. Paris"; Vol. I, pp. 359-364, L'Institut Textile de France: Paris, 1966.
28. Swift, J.A.; Holmes, A.W. Text. Res. J. 1965, 35, 1014-1019.
29. King, N.L.R.; Bradbury, J.H. Australian J. Biol. Sci. 1968, 21, 375-384.
30. Lofts, P.F.; Truter, E.V. J. Text. Inst. 1969, 60, 46-56.
31. Moncrieff, R.W. "Wool Shrinkage and Prevention"; National Trades Press: London, 1953.
32. Robbins, C.R. "Chemical and Physical Behaviour of Human Hair"; pp. 153-207, Litton Educational Publishing Inc: New York, 1979.
33. Makinson, K.R. "Shrinkproofing of Wool"; Marcel Dekker: New York, 1979.
34. Howell, H.; Mieszkis, K.; Tabor, D. "Friction in Textiles"; Butterworths: London, 1959.
35. Orwin, D.F.G.; Thomson, R.W.; Flower, N.E. J. Ultrastruct. Res. 1973, 45, 1-14.

36. Chapman, R.E.; Gemmell, R.T. J. Ultrastruct. Res. 1971, 36, 342-354.
37. Rogers, G.E. Ann. N.Y. Acad. Sci. 1959, 83, 378-399.
38. Kaplin, I.J.; Whiteley, K.J. Australian J. Biol. Sci. 1978, 31, 231-240.
39. Orwin, D.F.G.; Woods, J.L.; Elliott, K.H. in "Proc. 6th Int. Wool Text. Res. Conf. Pretoria"; Vol. II, pp. 193-205, South African Wool and Textile Research Institute: Pretoria, 1980.
40. Peters, D.E.; Bradbury, J.H. Australian J. Biol. Sci. 1976, 29, 43-55.
41. Elöd, E.; Zahn, H. Melliand Textilber. 1946, 27, 68-70.
42. Rogers, G.E.; Harding, H.W.J. in "Proc. 5th Int. Wool Text. Res. Conf. Aachen, 1975"; Vol. II, pp. 212-221, Deutches Wollforschungsinstitut: Aachen, 1976.
43. Harding, H.W.J.; Rogers, G.E. Biochemistry, 1971, 10, 624-630.
44. Fraser, R.D.B.; MacRae, T.P.; Millward, G.R.; Parry, D.A.D.; Suzuki, E.; Tulloch, P.A. Appl. Polymer Symp. 1971, 18, 65-83.
45. Woods, H.J. Proc. Roy. Soc. (Phys.) 1938, 166, 76-96.
46. Steinert, P.M.; Starger, J.M.; Goldman, R.D. in "Fibrous Proteins: Scientific, Industrial and Medical Aspects"; (Parry, D.A.D. and Creamer, L.K. eds.) Vol. 2, pp. 227-236, Academic Press: London, 1980.
47. Jones, L.N. Biochim. Biophys. Acta 1976, 446, 515-524.
48. Fraser, R.D.B.; MacRae, T.P.; Rogers, G.E. Nature (London) 1959, 183, 592-594.
49. Fraser, R.D.B.; Gillespie, J.M.; MacRae, T.P. Comp. Biochem. Physiol. 1973, 44B, 943-947.
50. Gillespie, J.M.; Frenkel, M.J. in "Proc. 5th Int. Wool Text. Res. Conf. Aachen, 1975"; Vol. II, pp. 265-284, Deutches Wollforschungsinstitut: Aachen, 1976.
51. Bradbury, J.H. Advan. Protein Chem. 1973, 27, 111-211.
52. Fraser, R.D.B.; MacRae, T.P. "Conformation in Fibrous Proteins"; Academic Press: New York, 1973.
53. Crick, F.H.C. Acta Cryst. 1953, 6, 689-697.
54. Lindley, H. in "Chemistry of Natural Protein Fibers"; (Asquith, R.S. ed.) pp. 147-191 Plenum Press: New York, 1977.
55. Crewther, W.G. in "Proc. 5th Int. Wool Text. Res. Conf. Aachen, 1975"; Vol. I. pp. 1-101 Deutches Wollforschungsinstitut: Aachen, 1976.
56. Suzuki, E.; Crewther, W.G.; Fraser, R.D.B.; MacRae, T.P.; McKern, N.M. J. Mol. Biol. 1973, 73, 275-278.
57. Gough, K.H.; Inglis, A.S.; Crewther, W.G. Biochem. J. 1978, 173, 373-385.
58. Crewther, W.G.; Inglis, A.S.; McKern, N.M. Biochem. J. 1978, 173, 365-371.
59. Crewther, W.G.; Dowling, L.M. Appl. Polymer Symp. 1971, 18, 147-156.
60. Fraser, R.D.B.; MacRae, T.P.; Miller, A. J. Mol. Biol. 1964, 10, 147-156.

61. Fraser, R.D.B.; MacRae, T.P.; Miller, A. J. Mol. Biol. 1965, 14, 432-442.
62. McLachlan, A.D. J. Mol. Biol. 1978, 124, 297-304.
63. Parry, D.A.D.; Crewther, W.G.; Fraser, R.D.B.; MacRae, T.P. J. Mol. Biol. 1977, 113, 449-454.
64. Parry, D.A.D. in "Fibrous Proteins: Scientific, Industrial and Medical Aspects"; (Parry, D.A.D. and Creamer, L.K. eds.) Vol. I, pp. 393-427 Academic Press: New York, 1979.
65. Crewther, W.G.; Dobb, M.G.; Dowling, L.M.; Harrap, B.S. in "Symposium on Fibrous Proteins, Australia 1967"; (Crewther, W.G. ed.) pp. 329-340 Butterworths: Australia, 1968.
66. Amiya, T.; Miyamoto, T.; Inagaki, H. Biopolymers 1980, 19, 1093-1097.
67. Elleman, T.C. Biochem. J. 1972, 128, 1229-1239.
68. Elleman, T.C. Biochem. J. 1972, 130, 833-845.
69. Elleman, T.C.; Dopheide, T.A. J. Biol. Chem. 1972, 247, 3900-3909.
70. Swart, L.S.; Haylett, T. Biochem. J. 1973, 133, 641-653.
71. Swart, L.S.; Joubert, F.J.; Parris, D. in "Proc. 5th Int. Wool Text. Res. Conf. Aachen, 1975"; Vol. II, pp. 254-263, Deutches Wollforschungsinstitut: Aachen, 1976.
72. Elleman, T.C.; Lindley, H.; Rowlands, R.J. Nature (London) 1973, 246, 530-531.
73. Swart, L.S. Nature (New Biol.) 1973, 243, 27-29.
74. Parry, D.A.D.; Fraser, R.D.B.; MacRae, T.P. Int. J. Biol. Macromol. 1979, 1, 17-22.
75. Smith, J.A.; Pease, L.G. Crit. Rev. Biochem. 1980, 8, 315-399.
76. Haylett, T.; Swart, L.S.; Parris, D. Appl. Polymer Symp. 1971, 18, 37-44.
77. Dopheide, T.A. Eur. J. Biochem. 1973, 34, 120-124.
78. Marshall, R.C.; Gillespie, J.M.; Inglis, A.S.; Frenkel, M.J. in "Proc. 6th Int. Wool Text. Res. Conf. Pretoria"; Vol. II, pp. 147-158, South African Wool and Textile Research Institute: Pretoria, 1980.
79. Whitmore, P.G. J. Cell Biol. 1972, 52, 174-179.
80. Jones, L.N. Biochim. Biophys. Acta. 1975, 412, 91-98.
81. Bendit, E.G. Textile Res. J. 1968, 38, 15-21.
82. Filshie, B.K.; Rogers, G.E. J. Mol. Biol. 1961, 3, 784-786.
83. Millward, G.R. J. Cell Biol, 1969, 42, 317-320.
84. Millward, G.R. J. Ultrastruct. Res. 1970, 31, 349-355.
85. Fraser, R.D.B.; MacRae, T.P.; Suzuki, E. J. Mol. Biol. 1976, 108, 435-452.
86. Crewther, W.G.; Dowling, L.M.; Inglis, A.S. in "Proc. 6th Int. Wool Text. Res. Conf. Pretoria"; Vol. II, pp. 79-91, South African Wool and Textile Research Institute: Pretoria, 1980.
87. Fraser, R.D.B.; MacRae, T.P. Biochim. Biophys. Acta 1958, 29, 229-240.

88. Leach, S.J.; Rogers, G.E.; Filshie, B.K. Arch. Biochem. Biophys. 1964, 105, 270-287.
89. Rogers, G.E.; Filshie, B.K. in "Ultrastructure of Protein Fibers"; pp. 123-138 (Borasky, R. ed.), Academic Press: New York, 1963.
90. Dobb, M.G. J. Textile Inst. 1970, 61, 232-236.
91. Dobb, M.G.; Sikorski, J. Appl. Polymer Symp. 1971, 18, 743-756.
92. MacArthur, I. Nature (London) 1943, 152, 38-41.
93. Bear, R.S.; Rugo, H.J. Ann. N.Y. Acad. Sci. 1951, 53, 627-648.
94. Fraser, R.D.B.; MacRae, T.P. Polymer 1973, 14, 61-67.
95. Ahmadi, B.; Boston, N.M.; Dobb, M.G.; Speakman, P.T. in "Fibrous Proteins: Scientific, Industrial and Medical Aspects"; (Parry, D.A.D. and Creamer, L.K. eds.) Vol. II, pp. 161-166, Academic Press: London, 1980.
96. Crewther, W.G.; Dowling, L.M.; Gruen, L.C.; Sparrow, L.G.; Woods, E.F. in "Proc. 6th Int. Wool Text. Res. Conf., Pretoria"; Vol. II, pp. 1-12, South African Wool and Textile Research Institute: Pretoria, 1980.
97. Fraser, R.D.B.; Jones, L.N.; MacRae, T.P.; Suzuki, E.; Tulloch, P.A. in "Proc. 6th Int. Wool Text. Res. Conf., Pretoria"; Vol. I, pp. 1-33, South African Wool and Textile Research Institute: Pretoria, 1980.
98. Steinert, P.M.; Idler, W.W.; Poirier, M.G.; Katoh, Y.; Stoner, G.D.; Yuspa, S.H. Biochim. Biophys. Acta. 1979, 577, 11-21.
99. Spei, M. in "Proc. 6th Int. Wool Text. Res. Conf., Pretoria"; Vol. II, pp. 263-272, South African Wool and Textile Research Institute: Pretoria, 1980.
100. Spei, M. Appl. Polymer Symp. 1971, 18, 659-662.
101. Wilson, G.R. Polymer 1972, 13, 63-68.
102. Wainwright, S.A.; Biggs, W.D.; Currey, J.D.; Gosline, J.M. "Mechanical Design in Organisms:" Wiley: New York, 1976.
103. Stam, P.B.; Kratz, R.F.; White, H.J. Text. Res. J. 1952, 22, 448-465.
104. Feughelman, M. Text. Res. J. 1959, 29, 223-228.
105. Spei, M.; Zahn, H. Melliand Textilber. 1979, 60, 523-525.
106. Bendit, E.G., in "Proc. 6th Int. Wool Text. Res. Conf. Pretoria"; Vol. II, pp. 247-262, South African Wool and Textile Research Institute: Pretoria, 1980.
107. Gillespie, J.M.; Marshall, R.C. in "Proc. 6th Int. Wool Text. Res. Conf. Pretoria"; Vol. II, pp. 67-77, South African Wool and Textile Research Institute: Pretoria, 1980.
108. Feughelman, M.; Watt, I.C. Text. Res. J. 1964, 34, 643-644.
109. Feughelman, M.; Druhala, M. in "Proc. 5th Int. Wool Text. Res. Conf. Aachen, 1975"; Vol. II, pp. 340-349, Deutches Wollforschungsinstitut: Aachen, 1976.
110. Bendit, E.G. in "Proc. 5th Int. Wool Text. Res. Conf. Aachen, 1975"; Vol. II, pp. 351-361, Deutches Wollforschungsinstitut: Aachen, 1976.

111. Bendit, E.G.; Gillespie, J.M. Biopolymers 1978, 17, 2743-2745.
112. Bendit, E.G. in "Fibrous Proteins: Scientific, Industrial and Medical Aspects"; (Parry, D.A.D. and Creamer, L.K. eds.) Vol. II, pp. 185-194, Academic Press: London, 1980.
113. Feughelman, M. Text. Res. J. 1979, 49, 704-710.
114. Schutz, R.A. ACS Symp. Ser. 1977, 50, 232-251.
115. Holmes, A.W. J. Soc. Cosmetic Chem. 1964, 15, 595-608.
116. Speakman, J.; Smith, S. J. Soc. Dyers Col. 1936, 52, 121-133.

RECEIVED March 18, 1982.

8

Man-Made Fibers: Science, Industry, and the Future

Richard Steele

Celanese Corporation, New York, NY 10036

The man-made fiber industry has matured and is entering a period of slow growth with a reduced level of scientific and technical development which will be directed by different social and economic forces from those of past years. It is possible that many end-uses now served by fibrous products will be taken over by laminar products based on new technology.

Man-made fibers had their origins some hundred and fifty years ago as natural historians became increasingly curious about the composition and structure of materials. In those victorian days such scientific curiosity was always linked to expectations that natural materials could be improved, made more useful for mankind or, eventually, be replaced by a man-made product.

Our world has been transformed from those different, but probably no simpler, days. Basic needs such as housing, communication and transportation have no doubt changed more since the middle of the nineteenth century than since the beginning of history. We have moved from a time when messages from one part of the world to another were carried by a messenger who traveled hours or weeks or years, to an era when most parts of the earth can be reached by telephone within a few minutes. In the early eighteen-hundreds locomotives were beginning to move people faster than sailing ships, that is, more than about 12 miles per hour, which had been the fastest means of travel for thousands of years. In our time the remarkable thing about the recent flight of the Columbia is not that it could travel around the world at 25 times the speed of sound, but that it is a vehicle which is expected to make a hundred such trips during its useful life. Inextricably enmeshed with developments like these have been proliferation and growth of all sorts of human activity. Natural history became natural science and then became a host of sciences which continue to multiply by what can only be called an

8412-0740-2/82-0139$6.00/0

atomization process. Along the way the interface of science and its useful application became known as technology. The evolution of science, technology and industry has taken place at the same time as an enormous increase in the world population, and it is inconceivable that the latter could have occurred in the absence of the former.

Why do I remind you of these things? It is because I think that the development of the science, technology, and industry of man-made fibers is prototypical of industrialization and growth processes, and that to understand where we are today vis-a-vis man-made fibers and, perhaps more importantly, what their future may be like, it is useful if not necessary to understand the whole sweep of their development.

Some of the points I want to make were suggested or reinforced as I reviewed the scientific and technical work of Milton Harris. His entire published work, with the exception of a few papers on cellulose acetate, deals with natural fibers, -- wool, silk and cotton. Yet no one who has worked in the synthetic fiber field over the past forty years would fail to acknowledge that the understanding of natural fibers achieved by Harris and his co-workers has been not only in many cases applicable to synthetics, but also a very considerable spur to much work in this field.

I would like to take up three ideas. The first is that expansion of the demand for fibers and the subsequent growth of their industrial production have been the key factors in the development of the science and technology of synthetic fibers. The second is that the present state of the industry and its prospects for growth suggest a qualitatively different period of development over the next decades. Finally, I would also suggest that the human needs that fibers have fulfilled will begin to be increasingly met by products of a different sort.

The Relation of Science to Industrial Growth

It is hard to recollect the nature of society at the beginning of the industrial age. After several centuries of an explosion of exploration, when the limits of the world had been probed, its lands and seas mapped, its plant and animal life described and classified, there was coming into being the sense of science as a rational understanding of the universe and as a guide to its exploitation by and for men and women. The realization of this perception was the industrial revolution. I would suggest that without the rapid application of this dawning understanding, through the industrial revolution, natural history would not have developed into the science that we know now.

Social and humanistic forces were involved also. Man's conception that there could be made--to get to our chief concern--fibers that would be better, more available (and therefore cheaper) and more specialized than wool or silk or cotton, was not unconnected to competition for profit and capital accumulation. Natural history and industrialization, what we now would more likely call science and technology, alternatively pushed and pulled at the myriad projects that optimistic victorians devised for the betterment of the human race and the lining of their own pockets.

Spinning frames and power looms were among the early inventions of the industrial revolution and initiated rapid growth in the demand for and consumption of fibers. Nations with positions based on the age of exploration grew rich providing American or Indian cotton or Australian wool to meet these demands. People began to think of making fibers in factories, particularly in countries without access to such resources. However, infant science poorly served the artisans, inventors, and technologists who tried to find ways to manufacture fibers. Cellulose was identified by Payen as a chemical substance in 1852. Its chemical structure was elaborated in the 1920's and its supermolecular morphology was reasonably well described 30 to 40 years later. But in this same span of time a world industry was established for manufacturing fibers of cellulose and it depended very little on the science of cellulose. Indeed, knowledge was sometimes used in a naively irrelevant way. For example, Chardonnet, trying to win an imperial prize for making an artificial silk that would free France's weaving trade of dependence on the Orient, chose to work with cellulose from mulberry trees because silkworms fed on mulberry leaves. He was unaware or uncaring of the chemical difference between cellulose and silk protein. Cellulose was a center of attention for many fiber makers although attempts, some successful, were made to transform many other natural raw materials into fibers. Nevertheless cellulose eventually became the basis of the dominant process for making man-made fibers by such transformations. The inventions of viscose by Cross and Bevan and of many mechanical elements key to the spinning process, such as the Topham box, were funded by the old English firm of silk-weavers, Samuel Courtauld and Son, who were spurred by the need for raw material for their insatiable mechanized looms. The first viscose plants were built about the turn of the century and the industrial firms who have dominated the rayon industry were in place and established by 1930. The world-wide industry and the basic science were more-or-less completed over the same stretch of time. In general, the industry was probably the leader and provided the incentive, and the money, for the scientific understanding that was simultaneously being achieved. The industry did not grow on a reservoir of knowledge, it grew by trial-and-error and/or science, while driven by essentially commercial demands.

A most important point to observe is that science and industrial technology, in whatever relationship they achieve, expand most rapidly during the prime growth and consolidation phase of an industry's development. As new plants are built, new solutions to technological problems are found. As a new product encounters broader end-uses, new technology for product modification must be developed. When industry expansion stops, science and technological development also stop. When the industrial growth has been stopped by the introduction of an intrinsically better technology or product, such as, in the case of viscose fiber, the melt-spun fibers nylon and polyester, the industry enters a period of decline or, at best, maturity. A few statistics show what has happened in the competition between viscose and polyester. In 1965, two hundred viscose plants yielded a little more than half of the world's man-made fiber production. Now half the number of plants produce one-fifth of world consumption. Polyester goes the other way, from a hundred plants to about 250, and from 9% of man-made fiber production to almost 40%. It is premature for polyester producers to gloat, however, because the period of their rapid growth is probably over and the positive interaction among plant expansion, technological development, and scientific understanding has begun to fail.

This is a typical industrial phenomenon when high relative growth stops or new technology intervenes. Other examples are the post-office vis-a-vis the telephone, the railroads versus highway transport, or the meat-packing industry after new marketing and distribution systems evolved. Neither government subsidy, basic research nor technological development will restore such industries to their vigorous youth. Nor will the man-made fiber industry be the same in the coming decades.

The Beginning of the Future

The man-made fiber industry will be different in the next decades in several important ways. The growth and change that have characterized it over the past 75 years will have a different quality. It will have a new relationship to science. The successful parts of the industry will be managed in a new style. It will not become as obsolete as the post office. It will do a better job of adapting to a new role than the railroads. It will find a continuing place in a steady-state industrialized world.

The previous section referred to many details of one of the sub-industries of fiber manufacturing, solution spinning. It matured earlier than the other main sub-industry, melt-spinning, but there are many fascinating parallels between the two. Their life cycles will be basically quite similar if not identical and some comparisons will be shown later. However, looking at the fiber industry as a whole, its initial growth phase, now ended, was based on two elements: an expansion of the per capita consumption of fibers, particularly in those industrializing nations who could afford it, and a displacement of the traditional natural fibers. During the next years growth could come from three factors: (1) population growth in the countries now using high per capita levels of fibers, (2) an increasing capability of less affluent nations, and (3) development of major new end-uses.

It is hard to conceive that the population of the world can continue to grow at the rate of the past hundred and fifty years. It is apparent that growth in industrialized, richer countries has slowed down and will probably continue to do so. Population growth in this part of the world will require debottlenecking, expansion and an occasional new plant to provide for future fiber needs. In the less-developed world there seems to be more liklihood that Malthusian predictions will come true. However population growth will continue more to constrain and stress the industrial and economic potential of these countries. In spite of their optimism and desire and of spots of successful rapid development, the capacity of the less-developed countries to afford and consume fibers is going to grow at a relatively slow rate. It will increase, and eventually it will reach a level appropriate to each country's well-being and life-style. However, it is not realistic to assume that this will occur at a rate similar to the growth rate of the fiber industry over the past 75 years.

Consider as an example the People's Republic of China. Among their original proposals for industrialization were two polyester plants which would have been the largest in the world and which would have produced about two pounds of fiber per year per citizen of China. These plants would have included units for making paraxylene and terephthalic acid as well as polymer and fiber. As China marshalls its resources and gathers it strength, it is clear that they can and will build plants of this appropriate scale. However, the Chinese themselves now suggest that these plants will better fit their plans sometime in the future than in the first half of the eighties. China's problem in this respect is a magnified example of what each under-industrialized country faces with respect to raising capital, obtaining technology and building infrastructures to obtain a sufficient fiber supply for their present and growing populations. The rate of achievement will be slow, even if world cooperation could be maximized in such an effort. Fiber manufacturing will grow slowly, under non-competitive conditions, and the stimulus for rapid technological changes will not be present in such growth.

Markets for man-made fibers have grown primarily by displacing natural fibers in traditional textile end-uses. The degree of penetration of such markets may increase somewhat, but there are no longer major end-uses--like sheetings or floor coverings--whose penetration or takeover could provide sudden growth. Somewhat different are markets which originate from an invention in other fields--for example, the pneumatic automobile tire--where a large-end use has developed. It is hard to predict whether there will be invented new technologies that will produce similar markets in the future. None seems to be on the horizon, but such discoveries are by their nature unforeseeable. In any event, it seems unlikely that the market for man-made fibers will expand significantly because new ways to use them have been discovered. (During the discussion of this paper, a questioner correctly observed that, in spite of my automobile tire example, I seemed to be thinking primarily of textile possibilities and not, for example, of the important area of fiber-reinforced composites.)

In summary, the man-made fiber industry is entering a period of mature growth and expansion quite unlike the rapid growth of earlier days. The driving forces for expansion will be (1) population growth in the industrialized world and (2) slowly increasing industrial and economic power in the less developed sector. The rate of growth from these factors will not provide a driving force for basic structural change in the industry. Present dominant technologies--viscose in solution spinning, polyester in melt spinning--will continue to be most widely used and most expansion will probably be in the latter.

If this relatively steady-state future is examined more closely, it becomes obvious that it will be a dynamic situation and that changes will occur, many the result of extrinsic forces. Plants and their supporting technologies wear out and must be replaced. Plants for making cellulose acetate, for example, now use computers to operate the technology that evolved in the period from 1890 to 1920. Environmental, energy cost pressures, and other external factors will also make reinvestment necessary, as well as continuing pressure for cost competitiveness and productivity.

To emphasize the different character of this scenario, consider some of the changes of past years. In solution spinning, viscose emerged as a commercial product in the early years of the century, but its dominance of this sub-industry was not established until many years later after it had been challenged by acetate and acrylic fibers and by products based on seaweed, soybeans, feathers and a host of other materials. In melt-spinning, the early frontrunner, nylon, failed to achieve dominance and was overtaken by polyester. Near-revolutionary effects could also result from far simpler events than the technological competition implied in the above examples. It was not until the later twenties, for example, that the idea of cutting "artificial silk" into staple fiber was conceived and within a few years the potential market for man-made fibers was multiplied many times. About the same time Dupont hired a scientist from Harvard to exploit Staudinger's avant-garde ideas of what polymers were and how they could be made. Within a few years nylon and melt-spinning had been developed. These were heroic years when new worlds of industry were being shaped.

But after an age of heroes comes a time of builders. The

emergence of a clear assessment of the intrinsic usefulness of fiber-making technologies (the assessment that can lead to the claim of dominance) and the consolidation of the industry around the groups committed to those dominant technologies provide a basis for the work of builders, who now have a structural organization in place, a reasonable view of where and what their markets will be, and a clear role in the world's industrial sector. The new directions, jostling, free-wheeling, and entrepreneurism that have characterized the fiber industry for so long are subsiding and it will be managed in quite a different way in the future.

Before discussing some of the parameters which will characterize successful management of the fiber business in the future, let us consider the continuing role of science and technology. The industrial revolution was a popular one because its success was seen by most people to bring jobs, wages and prosperity in greater measure than pre-industrial economies could achieve. Smokestacks and factories became symbols of vigorous, growing communities that were attractive because of the jobs and wages they promised for workers and their children. Successive generations reached higher standards of living than their fathers had known. This is the result that undeveloped countries now yearn for. Unfortunately they can now see this social transformation more wholly, with all its difficult and unhappy aspects, and they want not only to avoid these but to achieve the positive results in the shortest possible time. Past experience hasn't taught us how to do this, and whatever schemes are devised to try to achieve this utopian goal will probably involve difficulties and mistakes that will produce disappointing side-effects. As suggested earlier, these changes cannot take place rapidly.

However, most of the man-made fiber industry is in the industrialized countries and for the rest of this century will probably continue to be there. The social pressure on industry in such countries is now not for more jobs but for better jobs that leave more time for other interests. There is a popular yearning for a better quality of life--whatever that vague phrase means. Jobs that require a trade-off of environmental amenities, long hours or even hard work are no longer acceptable to many people. Science and technology will be important to industry, and to the man-made fiber industry in particular, to the degree that they can serve these changing expectations.

In the fiber industry this will mean that we should look for the following kinds of developments. First are those that improve the "human" aspects of a job, reducing the animal-like work involved. Running an acetate flake process through a computer rather than by hands-on control is an example where the mental component of a task is emphasized at the expense of physical effort. As suggested above, such a change may be triggered by plant obsolescence or for cost reasons, but the form it takes will have to be shaped by operator expectations of what a good job should be if a stable and productive work force is to continue to be attracted. Similarly, the way processes are carried out must change in directions that reduce elements of environmental stress--noise, health concerns, heat, pollution, etc. Laws and regulations that are forcing such moves are reflections of basic wants and, while they may be modified or slowed down, they will not be reversed and will become an accepted part of the fabric of society.

A third area where technical development will concentrate is productivity. This is not just because we're falling behind Japan, Germany or other industrial nations. More importantly, such concentration is needed to meet the social pressures outlined above with respect to the place that jobs will have in peoples lives, the nature of the work, and the resources they provide to enrich, or distract, the non-working time people increasingly have. Like other social pressures, this one generates economic pressure, obviously, and the latter will be very powerful force directing scientific and technical effort into the area of productivity. Productivity improvement is in general capital intensive and an economic return on such investment is required sooner or later by all accumulators of capital whether they take the form of corporations or socialist states.

Finally, there will continue to be new product development. There are not likely to be new fibers, except perhaps very specialized ones. Nor will there be major changes such as the introduction of high wet modulus rayon or draw-textured polyester filament, which opened up large new fields of application. But within the bounds of present product ranges there will be modifications and improvements. The successful ones will be those which contribute to the changing expectations discussed above. New fiber modifications will be accepted if they yield final fabrics, garments or other articles which consumers will recognize as contributing to their comfort, sense of well-being, and relaxed life-style. They will also be accepted if they help with the problems of reducing environmental stress, increasing productivity, or making a job easier.

Now the characterists can be outlined of the next stage of fiber industry-development which will determine the kind of management it will require:

1. Growth will continue but at a slower rate. The markets which in the past have supported proliferation of a wide range of fibers and fiber products will be growing much more slowly. The need for product modifications to penetrate new end-use areas will diminish. The continual investment in new plants which has facilitated process and product improvement will be reduced.

   The large fiber markets in the less developed countries will continue to grow but with social and economic restraints. The need of less developed countries to satisfy very basic needs is so great that emphasis will be on commodity products, so that growth will not be a stimulus to product innovation.

2. The pattern of technological dominance is clear. New plants for melt-spinning will manufacture polyester. If solution-spinning plants are needed--as a result of continuing cost increases in oil-based raw-materials--, they will make viscose rayon. Polyester and rayon will be the commodity fibers of the future, along with cotton, and will be sold without the merchandising of the past 50 years. Other fibers will continue to be produced,--nylon, wool, acetate, aramid, acrylic and so on,--but will be used to meet specialized performance or esthetic requirements.
3. Process technology will change in directions that take into account new worker expectations of what an industrial job should be like. These expectations imply themselves a need for productivity improvement, but the latter will be needed also to make the necessary changes economically possible.

The Discomfort Factor

The above conclusions about the future of man-made fibers are the result of examining the interaction of scientific, technological and industrial factors for the past century or so and interpreting them in light of social and economic forces important in the world today. It is a conventional projection of the past into the future and it leaves me uncomfortable. There is very probably going to be more excitement in the future than this picture suggests.

An alternative assumption that surely needs to be examined is that there will be fundamental changes in the way that the needs now satisfied by fibers will be met. For example, the materials used for clothing, household articles, or floor coverings might no longer involve weaving or knitting or tufting of fibers to form the sort of structures we are accustomed to. I cannot personally conceive of a structure, which might be like paper or film or nonwoven, which is likely to substitute satisfactorily for present materials. But my father would undoubtedly have had similar difficulties in conceptualizing jet propulsion or pictures sent through the air that move and talk. There seems to be a strong inhibition--a sort of future block--against accepting that fundamental and radical changes will continue to occur in the world.

I would like to turn back briefly to some of the industrial history mentioned before, particularly with respect to the two sub-industries: solution spinning and melt spinning. There are five stages of development that can be defined in each. The first is a period of conception, understanding, and gestation characterized by an interaction of science, commercial plans, and dreams, during which enough information is gathered that the desired end-result can be conceived of as feasible. In solution-spinning this was happening from 1850 to 1900. Beginning in the latter part of this period and continuing until 1920 or so very active commercial development occurred, with many new companies taking a chance on new products and processes as possibilities arose. It was during this time that viscose began to assume its dominant position. During the twenties the companies that were to be giants became well established and began a period of rapid growth that covered the next two decades, until the early fifties. After 15 or 20 years of maturity and slower growth, it was beginning to be obvious that melt-spinning, particularly of polyester, was beginning to gain a competitive edge over viscose and other solution spun fibers, and signs of their ultimate decline became obvious. Statistics given above demonstrate this changing pattern.

Melt-spinning has gone through similar stages at a somewhat later time. Its gestation was from 1910 to 1930; commercial development and the initiation of industrial activity covered 1930 to about 1960. Its period of great growth was 1960 to 1980. It is entering its decades of maturity. The uncomfortable question that arises is whether there will come to it a period of stagnation or decline.

This pattern of industrial change is apparent in many other sectors, as well as man-made fibers, and if it does not change as the world enters its mature industrialized phase (some call it a post-industrial phase), this pattern has a very important implication, which is that the gestation period, when the scientific and industrial possibilities are being explored and cultivated, occurs well before large scale commercial development, before full assessments of various technologies can be made, and certainly before much scientific understanding has been developed. If this pattern holds for the present situation in the man-made fiber industry, it suggests that the gestation period of its successor is occurring now, that the bits and pieces of appropriate technology are being worked out and that possible exploitation of relevant scientific discoveries is being explored. Indeed, the size and success of the fiber industry may itself provide a strong incentive for the development of an alternate industry.

The problem for those who would like to continue in this industrial sector is to identify what is happening in the scientific, technical and industrial areas that is relevant, and to move into a position to capitalize on appropriate commercial development when the time is right. It is desirable to establish a position soon enough not to be excluded from the dominant technology.

To the degree that fibers are used to make laminar materials (e.g., fabrics) suitable for apparel and household end-uses, one would guess that the relevant developments and discoveries are being made on the interface between the conventional textile industry and the paper, film and nonwoven industries. One can imagine that there could evolve from this area a novel process for making new kinds of laminar products which could more affectively compete for the end-use markets of textiles than the products we know now in these non-textile industries. It might even be as simple as a machine to which polymer is fed, perhaps polypropylene, and from which comes a kind of grey goods that can be adapted and finished to many end-uses. The industry must be watchful of developments in this area and must not underrate them. Products will improve, consumers standards of product acceptability and desirability will change, and textile markets will inevitably be eroded or usurped by such developments.

Conclusion

In summary, let me restate the propositions I have put forward:

First is that the history of the man-made fiber industry until the present has been prototypical industrialization, where rapid growth and expansion have driven and been driven by a growing demand for the industrial product. As this phase ends, the driving force of growth itself disappears and the need for technical and scientific innovations is much reduced.

The second is that industry will grow more slowly and the the requirements for technical or scientific input will be based in large measure on different social and economic forces from those most important in the past.

And finally, it is suggested that industries not only mature but eventually retrogress and that the man-made fiber industry, melt-spinning as well as solution-spinning, may fit into such a pattern. The new industry whose aggessive competition may cause this decline is in a period of gestation now. Many of its key elements are probably known and many people are trying to put them together.

The man-made fiber industry of the future will belong to those who recognize the changes maturity has brought, manage it well in its middle age, and best prepare for the emergence of a new industry capable of making products for many of the markets fibrous textiles have today.

RECEIVED July 18, 1981.

9

International Cooperation in Information Science

J. W. Barrett

R. H. Cole Limited, Purley, Surrey, CR2 2UY, England

This paper deals with the importance of information and its transfer to the advance and application of science, the need of treating information as a constantly expanding resource and the place of learned societies and institutions in the dissemination of scientific information. The problems of establishing a comprehensive global system for gathering and distributing chemical information are described and the vital importance of CAS to any such system emphasised. Above all appreciation is expressed of the major contributions of Milton Harris to promoting CAS as both a national and international force.

The chemist like all scientists advances his personal knowledge and understanding of his science in two main ways:

1. By observation and experimentation
2. By communication with other scientists

To establish a base from which he can advance, the chemist needs to know what is already known; this knowledge comes to him through communication of information, verbally from his tutors and peers as well as in written and other forms. As the complement of this receipt of information from outside and its conversion into personal knowledge, unless his sole objective is self-gratification, he will wish to pass on to others the results of his own observations and experimentation, hopefully enriched by his personal insights, so as to extend the universe of what is known and hence what can in future become known.

It may not be too extreme a position to assert that of the two major components of scientific activity, that is doing on the one hand and telling on the other, the act of communicating information is the more important though of course neither can exist in a real sense without the other. The making of scientific observations in the absence of a means of informing others of them would be a singularly sterile

8412-0740-2/82-0151$6.00/0

undertaking perhaps comparable in value and quality to masturbation as an aid to procreation.

That the transfer of information between scientists is a corner-stone, perhaps even the key-stone, in the conduct of science has long been recognised. The foundation by Charles II of the Royal Society in London was an early and in hindsight crucial recognition of this. Information is the feedstock for, though not the same thing as, knowledge. Without information and the mechanisms for transferring it freely amongst scientists there would be neither science nor scientific knowledge. In this sense scientific information is a raw material, a natural or rather human resource with some of the same qualities as the more accepted material resources such as coal, oil, ironstone. Without coal or oil there would be no petrochemical industry, without ironstone no steel and precious little engineering, without scientific information there would be no science based industry. Not only is scientific information and the transmission thereof essential for the promotion of science itself but also it is vital to the advance of all technologically based industry.

However, unlike most other raw materials, information has the potential not only of self-replenishment but of growth. The more that information is consumed the more there is of information, but of course this characteristic is but a potential which can only be realised if mechanisms are created, nurtured and developed to ensure the collection of information, its organisation and its dissemination.

Perhaps the analogy of information as a raw material can be pressed with advantage in comparison with trees. Trees in their own right have an aesthetic value, something to be enjoyed. Scientific information in its own right is satisfying intellectually but neither tree or information is of material advantage until it is harvested, transferred to others and used often in conjunction with other resources to produce products of value to man.

The recognition with the foundation of the Royal Society of the pre-eminence of communication as the single universal tool of science has continued with the formation and growth of learned societies across science and throughout the world. Chemistry has been especially well served by institutions such as the American Chemical Society and the Societies of Chemistry in the United Kingdom and Germany but there is a wide variation in the capabilities and organisation of national chemistry societies. The role they play in the dissemination of chemical information is certainly of a wide spread. It includes publications of books and journals, creation and management of systems for the abstracting and indexing of chemical literature, organisation of conferences, symposia and the like with varying degrees of emphasis. Commercial and publishing organisations have for long been also involved in the dissemination of chemical and other scientific information and in some countries governments are much concerned. Overall therefore the world network for the gathering, publishing and disseminating chemical and scientific information is complex and isomorphic, if not amorphous.

It is of course an illusion to suppose that institutions of themselves serve any useful purpose. Just as chemistry has been well served by its learned institutions so these institutions have been well served by the individual chemists whose visions and drive to achieve these visions are so well illustrated by Milton Harris. Almost to the month 50 years ago he became a CA abstractor and continued as that for 17 years, only to continue for another 12 years as one of the most outstanding section editors for Chemical Abstracts. It would take a long time to catalogue all his other contributions to information gathering and dissemination but there are two major areas of effort which I must comment on. Milton recognised the central importance of the Chemical Abstract Service system in organising and managing the free flow of information between chemists and to this end he worked untiringly to develop the unique capabilities of the group in Columbus, Ohio and to ensure its continued existence as an organisation independent of government and managed by chemists, on behalf of chemists and of users of chemistry. Beginning in 1962 Milton was advisor and consultant Executive Office of President, Office of Science and Technology and worked successfully to overcome the threat of takeover of scientific and technical information programmes in the USA by the Department of Defence. He helped to negotiate the first major contract for CAS from the National Science Foundation and advised with great wisdom the government and government agencies on problems of scientific and technical information. These activities coupled with his work within the ACS on behalf of CAS would be sufficient to guarantee Milton Harris a place of honour.

But it may be that it is in another area that we shall in hindsight see his major contribution. Beginning in 1964 Milton Harris worked towards a conception of an international chemical information system based around the CAS facility in Columbus, Ohio but giving a real opportunity of users of chemical information throughout the world, to have direct involvement in and responsiblity for the way in which the CAS international chemical information system would develop. His personal vision and enthusiasm spilled off to some outside the USA and we in the Chemical Society were spurred on to make an agreement with the American Chemical Society to permit us in 1966 to establish a research unit dedicated to examining ways in which the new methods being developed by CAS would be best used in the UK and maybe elsewhere. In 1968/69 he led the negotiations on behalf of ACS that culminated in marketing agreements between ACS, the Chemical Society and the German Chemical Society from which developed the network of international relationships that today makes the chemical information scene unique in science. We are often reminded that more than 75% of the information gathered by Chemical Abstracts comes from outside the United States and that some 65% of subscribers are outside this country. Multinational industry has not found one right way to handle a global resource for purchasing and supply to a global market. In general however the resource is harvested through regional or national units as is also the distribution of the final products. Milton Harris' great experience in international business as well as international science coupled with his belief in the universal availability of timely and effective scientific information systems as a keystone of chemistry and of its application enabled him to give great leadership to ACS and its international partners in chemical information for the creation and maintenance of a multinational chemical information operation based solidly on the Chemical Abstract service in Columbus. CAS can gather its abstracts from individual chemists of many countries, from other national chemical information centres or from its own literature analysts. The pattern has changed and can change further. CAS can distribute its information to individual users worldwide either directly or through national information centres - again the pattern is a changing one. A review of progress made was given by Russell Rowlett editor of CAS in 1979 which summarises the parts then being played by France, Germany, Japan, Switzerland and the UK. It is clear that the roles being played differ from country to country which is not surprising because of the different situations of learned societies, of national information systems, of government attitudes and of political views. New techniques such as on-line information transfer introduce new opportunities and new problems but much patience and persistence will be needed before an international chemical information system can be fully developed.

Chemical information alone is not sufficient to promote the growth of chemistry and its full application; one of our early ambitions was to bring together the world information on chemistry, on physics, on biology, on mathematics, on the separate parts of engineering etc,, so that the end user whether he were scientist or engineer, academic or industrialist, from institute or government, was able to gain all scientific and technical information pertinent to his particular need. On-line transfer using a variety of computerised information bases is going some way to satisfy this need but we are long way from a low cost, fully effective, internationally available system. We admire the efforts of the American Chemical Society to collaborate with other societies and information institutions and recognise that science is fortunate in having an International Council of Scientific Unions associated with which there is an Abstracting Board which brings together the abstracting and indexing institutions for separate disciplines on an international basis. However, inspiration and drive does not seem to come too easily from international groups of this nature. Can it come from national academies, it would be suspect if it came from a learned society of a single discipline. It looks to me, Milton that we still need you.

I and my colleagues in the UK join with all your colleagues here to thank you for the great contributions you have made right up to your 75th birthday and to ask you to keep it up.

RECEIVED July 8, 1981.

10

Recent Advances in Polymer Chemistry

C. G. Overberger

University of Michigan, Department of Chemistry, Ann Arbor, MI 48109

Two areas of my current research program will be discussed: 1) Hydrophilic polymers with conformational order in aqueous solution, and 2) synthesis, polymerization, and conformational studies of substituted polyprolines.

Synthetic Macromolecules as Therapeutic Agents

Work is under way on the synthesis of new macromolecules, both polyanions and polycations, which could be expected to be effective antitumor or antiviral agents, interferon inducers or in some other manner therapeutically useful. In particular, we wish to determine the effect of macromolecular structure on biological activity.

Three polyethylenimine backbone polymers with amino acid grafts, prepared in our laboratory, have shown activities which appear to be a function of the length of the polymer chain as well as whether the polymer is linear or branched. Specifically, the graft copolymer with histidine and serine on "high" molecular weight linear polyethylenimine ($[\eta]^{25°C}_{MeOH} = 0.244$) with approximately the same mole percent of histidine and serine showed less activity. A third sample made from branched polyethylenimine, of higher molecular weight than either of the linear polymers and with analogous histidine and serine grafts, showed activity comparable to the lower molecular weight linear polymer. We are currently exploring this result.

We have synthesized a highly ordered structure derived from hydrophilic backbone (linear polyethylenimine) with an asymmetric center directly attached to a nucleic acid base (structures 1a,b, 2, 3). Our nucleic acid models appear to closely resemble Poly U and Poly A. Significant hypochromicity is observed in all of our systems. Structures 1a,b were examined by means of circular dichroism, the first nucleic acid models without a phosphate-ribose backbone which could be so examined.

8412-0740-2/82-0157$6.00/0

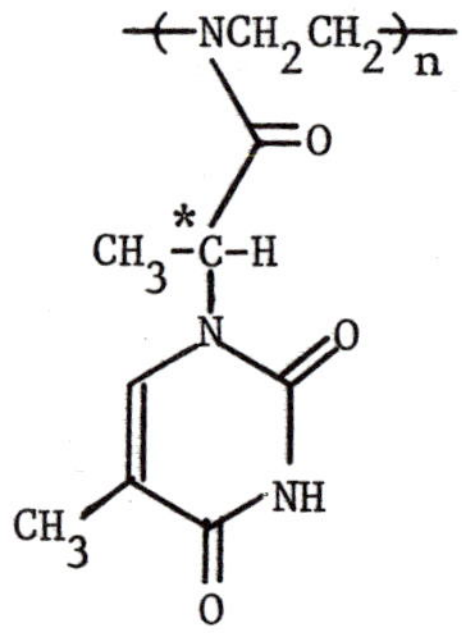

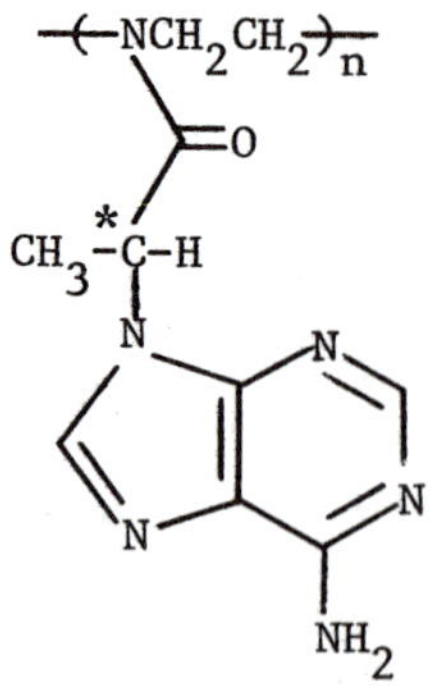

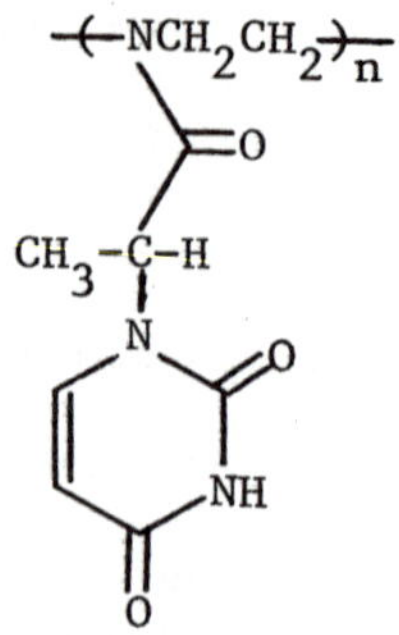

1 a) P-(−)-T

$[\alpha]_D^{25} = -48.4^\circ$

b) P-(+)-T

$[\alpha]_D^{25} = +49.5^\circ$

2

3

We are currently attempting to resolve compound 2 into its optical isomers for further studies by circular dichroism. We are also synthesizing a model for Poly C *via* the cytosyl pendant group. Modification of the structure of the macromolecule is initiated in two directions. In an attempt to further restrict the motions of the pendant group, a large alkyl substituent is placed on the chiral center. Also, higher molecular weight polyethylenimine as well as polyvinylamine are being used for the backbone. These modifications are made to increase our understanding of the system and hopefully its potential therapeutic value.

Synthesis, Polymerization and Conformational Studies of Substituted Polyprolines

The resolution of *cis*-5-ethylproline was successfully carried out (Scheme I).

Polymerization *via* N-carboxyanhydride of the *cis*-5-ethyl-D-proline was carried out in chloroform initiated with sodium methoxide. The polymer obtained is not soluble in alcohol, however it can be dissolved in trifluoroethanol and methanol mixed solvent. It is very soluble in chloroform and strong acids.

The conformation of poly(*cis*-5-ethyl-D-proline) (PC5EP) in solution has been studied by different spectroscopic techniques. In trifluoroethanol plus methanol, PC5EP gives a strong positive Cotton effect at 209 nm and a weak negative Cotton effect at 190 nm, hereto referred to as PC5EP-II; however, in tetrahydrofuran plus *n*-butanol PC5EP gives a strong negative Cotton effect at 214 nm and two positive Cotton effects at 230 nm and 198 nm (PC5EP-I). This result suggests PC5EP may assume two different ordered conformations, and in THF + *n*-BuOH, PC5EP behaves very similarly to that of poly-L-proline, Form I, since PP-Form I has a very similar CD spectrum with a very strong positive Cotton effect at 215 nm and two weak negative Cotton effects at 232 nm and 199 nm. In TFE + MeOH the general feature of the CD spectrum of PC5EP is also similar to that of PP-Form II; PP-Form II has a slightly blue shift (about 2 nm) of the peak at 207 nm and an additional peak at 226 nm.

A general scheme for the synthesis of *cis*- and *trans*-alkylprolines has been developed (Scheme II).

The separation procedure, which was used for 5-ethylproline, does not give satisfactory results in the 3-ethylproline case. The probable reason is that the ethyl group in the 5-position has a stronger influence on the formation of the sulfonamide than in the 3-position. Thus, when *cis*- and *trans*-3-ethylproline is reacted with *p*-toluenesulfonyl chloride, more sulfonamide is

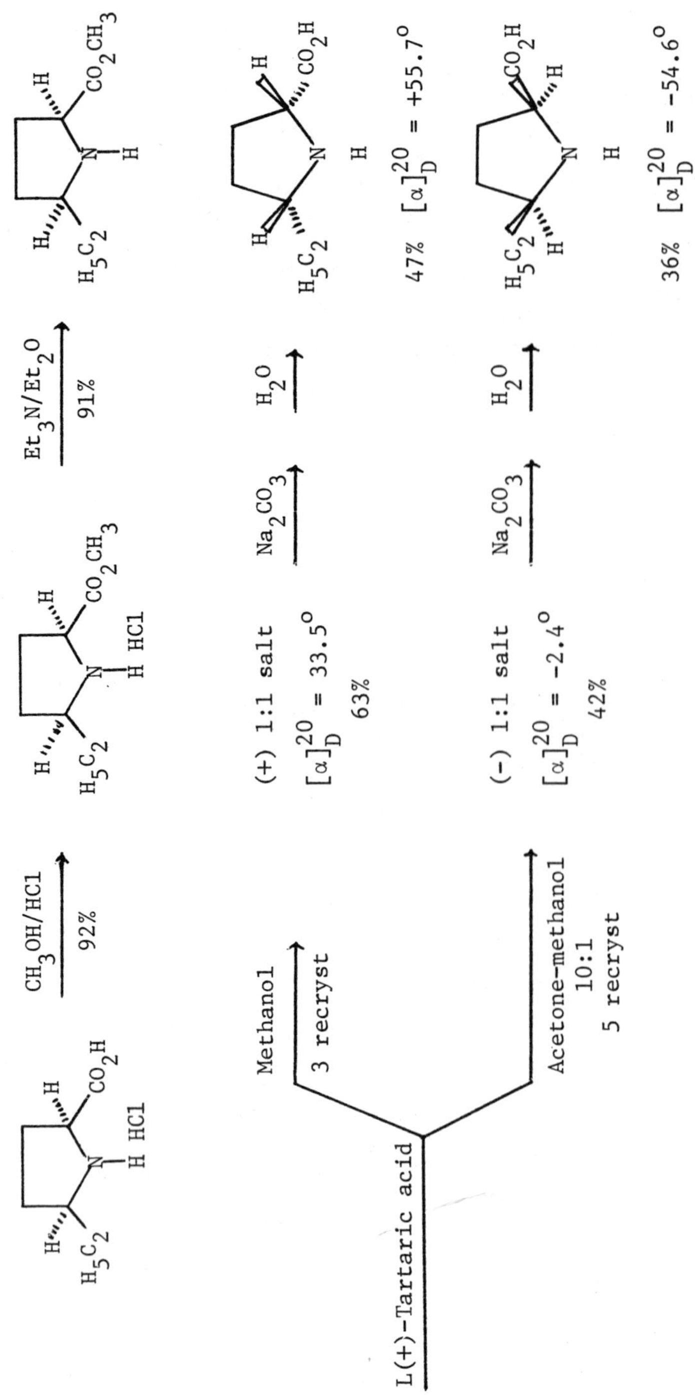

Scheme I. Resolution of cis-5-Ethylproline

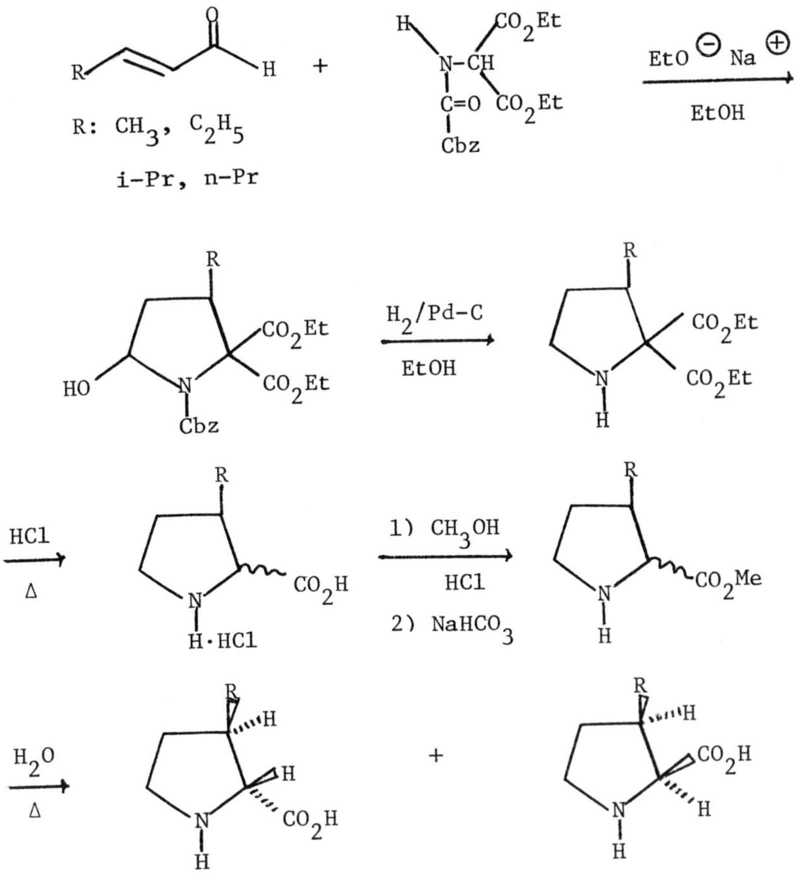

Scheme II. Synthesis of 3-Alkyl Prolines

obtained than expected, which suggests that the *cis*- and a large portion of the *trans*- have reacted. Currently we are working out a procedure which will transform the *cis*- and *trans*- mixture in the *cis*-isomer only.

Acknowledgment

A portion of this research was supported by the Division of Materials Research, Polymer Section, of the National Science Foundation.

RECEIVED June 25, 1982.

INDEX

D

I

K

L

M

T

U

V

W

X

Y

Z